A Course in Interacting Particle Systems

The culmination of years of teaching experience, this book provides a modern introduction to the mathematical theory of interacting particle systems. Assuming a background in probability and measure theory, it has been designed to support a one-semester course at a master's or PhD level. It also provides a useful reference for researchers, containing several results that have not appeared in print in this form before. An emphasis is placed on graphical representations, which are used to give a construction that is intuitively easier to grasp than the traditional generator approach. Also included is an extensive look at duality theory, along with discussions of mean-field methods, phase transitions, and critical behavior. The text is illustrated with the results of numerical simulations and features exercises in every chapter. The theory is demonstrated on a range of models, reflecting the modern state of the subject and highlighting the scope of possible applications.

JAN M. SWART is a research fellow at the Institute of Information Theory and Automation of the Czech Academy of Sciences, Prague. He has coauthored over 40 papers in probability theory with an emphasis on interacting particle systems and is known, among other things, for his work on the Brownian net.

CAMBRIDGE SERIES IN STATISTICAL AND PROBABILISTIC MATHEMATICS

This series of high-quality upper-division textbooks and expository monographs covers all aspects of stochastic applicable mathematics. The topics range from pure and applied statistics to probability theory, operations research, optimization, and mathematical programming. The books contain clear presentations of new developments in the field and also of the state of the art in classical methods. While emphasizing rigorous treatment of theoretical methods, the books also contain applications and discussions of new techniques made possible by advances in computational practice.

A complete list of books in the series can be found at www.cambridge.org/statistics.
Recent titles include the following:

38. *Nonparametric Estimation under Shape Constraints*, by Piet Groeneboom and Geurt Jongbloed
39. *Large Sample Covariance Matrices and High-Dimensional Data Analysis,* by Jianfeng Yao, Shurong Zheng and Zhidong Bai
40. *Mathematical Foundations of Infinite-Dimensional Statistical Models,* by Evarist Giné and Richard Nickl
41. *Confidence, Likelihood, Probability*, by Tore Schweder and Nils Lid Hjort
42. *Probability on Trees and Networks*, by Russell Lyons and Yuval Peres
43. *Random Graphs and Complex Networks* (*Volume 1*), by Remco van der Hofstad
44. *Fundamentals of Nonparametric Bayesian Inference*, by Subhashis Ghosal and Aad van der Vaart
45. *Long-Range Dependence and Self-Similarity*, by Vladas Pipiras and Murad S. Taqqu
46. *Predictive Statistics*, by Bertrand S. Clarke and Jennifer L. Clarke
47. *High-Dimensional Probability,* by Roman Vershynin
48. *High-Dimensional Statistics,* by Martin J. Wainwright
49. *Probability: Theory and Examples* (*Fifth Edition*), by Rick Durrett
50. *Model-based Clustering and Classification for Data Science,* by Charles Bouveyron *et al.*
51. *Spectral Analysis for Univariate Time Series*, by Donald B. Percival and Andrew T. Walden
52. *Statistical Hypothesis Testing in Context*, by Michael P. Fay and Erica H. Brittain
53. *The Fundamentals of Heavy Tails,* by Jayakrishnan Nair, Adam Wierman and Bert Zwart
54. *Random Graphs and Complex Networks* (*Volume 2*), by Remco van der Hofstad
55. *Modern Discrete Probability*, by Sébastien Roch
56. *Generalized Additive Models for Location*, Scale and Shape, by Mikis D. Stasinopoulos et al.
57. *Networks: Probability and Statistics*, by A. D. Barbour and Gesine Reinert
58. *High-Dimensional Probability (Second Edition)*, by Roman Vershynin

A Course in Interacting Particle Systems

Jan M. Swart
The Czech Academy of Sciences, Institute of Information Theory and Automation

CAMBRIDGE
UNIVERSITY PRESS

CAMBRIDGE
UNIVERSITY PRESS

Shaftesbury Road, Cambridge CB2 8EA, United Kingdom

One Liberty Plaza, 20th Floor, New York, NY 10006, USA

477 Williamstown Road, Port Melbourne, VIC 3207, Australia

314–321, 3rd Floor, Plot 3, Splendor Forum, Jasola District Centre,
New Delhi – 110025, India

Cambridge University Press is part of Cambridge University Press & Assessment,
a department of the University of Cambridge.

We share the University's mission to contribute to society through the pursuit of
education, learning and research at the highest international levels of excellence.

www.cambridge.org
Information on this title: www.cambridge.org/9781009843478
DOI: 10.1017/9781009843485

When citing this work, please include a reference to the DOI 10.1017/9781009843485

First published 2026

A catalogue record for this publication is available from the British Library

A Cataloging-in-Publication data record for this book is available from the Library of Congress

ISBN 978-1-009-84347-8 Hardback

Everyone has the right to life, liberty and the security of person.
Universal Declaration of Human Rights, Article 3.

This handsome volume provides an accessible introduction to one of the gems of modern probability theory. It will be valuable to those seeking a rigorous and well-illustrated account of the foundations and basic properties of Interacting Particle Systems.

Geoffrey Grimmett, University of Cambridge

This book on interacting particle systems is unique, developed as a textbook for a one-semester graduate course. Technicalities are kept to a minimum, making the treatment more accessible. The reader will receive a thorough and enjoyable education about the most important models, techniques, and results.

Rick Durrett, Duke University

This book serves as an important introduction to interacting particle systems, covering key foundational topics such as phase transitions, duality, and mean-field limits. The exposition is clear and well-written, with numerous examples that make the material engaging and accessible. It is particularly well-suited for Master's and PhD students, and I will certainly recommend it to my future students and research fellows.

Patricia Gonçalves, Instituto Superior Técnico

A Course in Interacting Particle Systems covers the construction, behavior, and modern tools of the most important particle systems: spin models; voter, contact, and exclusion processes. It displays great intuition combined with rigorous treatment and offers an outlook to current research. Numerous exercises help the journey in a novel way – a must-have read from a top expert in the field.

Márton Balázs, University of Bristol

A much-needed introduction to the field, at once accessible to beginners, and informative for experienced researchers. Intuitive and inviting, it focuses on Poisson graphical constructions and features richly illustrated descriptions of model behavior, while maintaining mathematical rigor with original formalism and clear proofs.

Daniel Valesin, University of Warwick

A bright and modern introduction to interacting particle systems. Both highly readable and mathematically rigorous, it presents key tools illustrated through a range of models. Complete with exercises, it offers a comprehensive and accessible entry point to this flourishing field.

Cristina Toninelli, University Paris Dauphine – PSL

Contents

Preface

Interacting particle systems, in the sense we will be using the word in this book, are countable systems of locally interacting Markov processes. Each interacting particle system is defined on a lattice: a countable set with (usually) some concept of distance defined on it; the canonical choice is the d-dimensional integer lattice $\mathbb{Z}^d$. Situated on each point in this lattice, there is a continuous-time Markov process with a finite state space (often even of cardinality two) whose jump rates depend on the states of the Markov processes on nearby sites. Interacting particle systems are often used as extremely simplified "toy models" for stochastic phenomena that involve a spatial structure.

An attractive property of interacting particle systems is that they are easy to simulate on a computer.[1] Although the definition of an interacting particle system often looks very simple, and problems of existence and uniqueness have long been settled, it is often surprisingly difficult to prove anything nontrivial about its behavior. With a few exceptions, explicit calculations tend not to be feasible, so one has to be satisfied with qualitative statements and some explicit bounds. Despite intensive research over more than 50 years, some easy-to-formulate problems still remain open, while the solutions of others have required the development of nontrivial and complicated techniques.

Luckily, as a reward for all this, it turns out that despite their simple rules, interacting particle systems are often remarkably subtle models that capture the sort of phenomena one is interested in much better than might initially be expected. Thus, while it may seem outrageous to assume that "Plants of a certain type occupy points in the square lattice $\mathbb{Z}^2$, live for an exponential time with mean one, and place seeds on unoccupied neighboring sites with rate λ," it turns out that making the model more realistic often does not change much in its overall behavior. Indeed, there is a general philosophy in the field, that is still insufficiently understood, which says that interacting particle systems come in "universality classes" with the property that all models in one class have roughly the same behavior.

As a mathematical discipline, the subject of interacting particle systems is still relatively young. It started around 1970 with the work of F. Spitzer [Spi70] and R. L. Dobrushin [Dob71], with many other authors joining in during the next few years. By 1975, general existence and uniqueness questions had been settled, four classical models had been introduced (the exclusion process, the stochastic Ising model, the voter model, and the contact process), and elementary (and less elementary) properties of these models had been proved. In 1985, when Liggett published his famous book [Lig85], the subject had established itself

[1] To get started doing this yourself, look at my simulation library that is available from `http://staff.utia.cas.cz/swart/simulate.html`.

as a mature field of study. Since then, it has continued to grow rapidly, to the point where it is impossible to accurately capture the state of the art in a single book. Liggett's second book [Lig99] focuses on three of the four classical models only. Such is the sophistication of modern methods that by now it would be possible to write a book on each of the four classical models alone.

While interacting particle systems, in the narrow sense defined earlier, have apparently not been the subject of mathematical study before 1970, the subject has close links to some problems that are considerably older. In particular, the Ising model (without time evolution) has been studied since 1925, while both the Ising model and the contact process have close connections to percolation, which has been studied since the late 1950s. In recent years, more links between interacting particle systems and other, older subjects of mathematical research have been established, and the field continues to receive new impulses not only from the applied, but also from the more theoretical side. Until 1990 most of the work concentrated on the d-dimensional integer lattice. Since then other lattices such as trees have gained popularity. Interacting particle systems on random graphs are a hot topic. Scaling limits, both deterministic (hydrodynamic limits) and random (Stochastic partial differential equations, super-Brownian motion, and the Brownian web), are an old subject that continues to see exciting developments.

Apart from Liggett's books [Lig85, Lig99], there exist a number of other books that treat interacting particle systems in one way or another. Durrett's lecture notes from 1988 [Dur88] and his St. Flour lecture notes [Dur95] still make great reading. Both include many pictures of numerical simulations, discuss the mean-field limit, and use percolation theory as a unifying idea behind the proofs. While his 1988 book focuses mainly on the classical models, his St. Flour lecture notes show how block arguments together with assumptions such as sufficiently long range interaction or rapid stirring can be used to treat a plethora of models. Both books focus on developing the great ideas and getting one's hands dirty doing actual calculations, as opposed to carefully developing the basic technical framework. The St. Flour lecture notes are quite high level, so Master students and even beginning PhD students may find it hard to see that all the technical details can really be filled in.

Two books that don't have interacting particle systems as their primary topic, but nevertheless say something useful about them, are Liggett's *Continuous Time Markov Processes* [Lig10] and Grimmett's *Probability on Graphs* [Gri18]. Apart from giving a short introduction to particle systems, Liggett's book is also a useful reference for the classical theory of continuous-time Markov chains. Grimmett's book includes a lot of material on percolation theory, which includes the contact process (as a form of oriented percolation) and the random cluster model with its application to the Ising and Potts models. Recently, Lanchier [Lan24] has made an impressive attempt to give a complete overview of the existing literature on interacting particle systems in the life and social sciences, which he managed at the cost of being only able to sketch the main proof ideas.

The present book grew out of lecture notes I wrote for courses I have been giving at Charles University in Prague at regular intervals starting in 2009. Since the idea was that it should be possible to cover most of the material in a one-semester course, a lot of basic material that can be found in Liggett's classic book [Lig85] is omitted here. What is new in the book has less to do with new results but more to do with how the material is presented. The introductory Chapter 1 shows a wide variety of models that reflect the present state of the subject. There is an informal discussion of phase transitions and of critical behavior, which is still poorly understood from a mathematical side, but nevertheless important to get the full picture. And there is a whole chapter (Chapter 3) devoted to the mean-field limit, which from the mathematical techniques involved is a bit off topic, but nevertheless essential to get a good complete understanding.

One of the most fundamental novelties is how graphical representations are given a central role in the construction of all kinds of interacting particle systems, instead of being viewed as a tool for the study of certain specific models only. This reflects the modern state of the art that indeed uses graphical representations all of the time. It also allows the basic existence and uniqueness results (presented in Chapter 4) to be proved in a way that prepares for the discussion of duality in Chapter 6. The basic existence and uniqueness result (Theorem 4.14) is a pathwise result, that despite being based on well-known methods, has not appeared in print in this form before. Most of the duality in this book is pathwise duality (which is a modern word for an old concept). Stochastic flows, both forward and backward in time, are given a central place.

The material is meant to be presentable (with minor omissions here and there) during a one-semester course. As a preparation for the book, the students need a basic course in measure theory and probability. It is also preferable if they have at least some prior experience with continuous-time Markov chains, so that some of the more standard sections of Chapter 2 can be skipped over quickly in favor of sections covering material that is less widely known. I have tried to make the book reasonably self-contained, but not at all costs, so material about differential equations or about Feller semigroups is cited without proof. Chapter 2 and Sections 4.1–4.5 describe the core technical results that one needs in order to understand the rest of the book. Chapters 3, and 5–7 have been written in such a way that they do not depend too much on each other mutually.

People that I am indebted to for their comments and suggestions include Tibor Mach, Aernout van Enter, Sam Olesker-Taylor, Jan Niklas Latz, Jim Fill, Cristina Toninelli, and I am sure more people whom I am now forgetting.

Work on this book was sponsored by Grant Agency of the Czech Republic (GAČR) grant 25-16267S.

1

Introduction

1.1 General Setup

Let S be a finite set, called the *local state space*, and let Λ be a countable set, called the *lattice*. We let S^Λ denote the Cartesian product space of Λ copies of S, that is, elements x of S^Λ are of the form

$$x = \big(x(i)\big)_{i\in\Lambda} \quad \text{with} \quad x(i) \in S \;\forall\, i \in \Lambda.$$

Equivalently, S^Λ is nothing else than the set of all functions $x\colon \Lambda \to S$.

Interacting particle systems are continuous-time Markov processes $X = (X_t)_{t\geq 0}$ with a state space of the form S^Λ. Thus, $(X_t)_{t\geq 0}$ is a Markov process such that at each time $t \geq 0$, the state of X is of the form

$$X_t = \big(X_t(i)\big)_{i\in\Lambda} \quad \text{with} \quad X_t(i) \in S \;\forall\, i \in \Lambda.$$

We call $X_t(i)$ the *local state* of X at time t and position i. Positions $i \in \Lambda$ are also often called *sites*.

The time evolution of continuous-time Markov processes is usually characterized by their *generator* G, which is an operator acting on functions $f\colon \mathcal{S} \to \mathbb{R}$, where $\mathcal{S}$ is the state space. For example, in the case of Brownian motion, the state space is $\mathcal{S} = \mathbb{R}$ and the generator is the differential operator $G = \frac{1}{2}\frac{\partial^2}{\partial x^2}$. In the case of an interacting particle system, the state space is of the form $\mathcal{S} = S^\Lambda$ and the generator can usually be written in the form

$$Gf(x) = \sum_{m\in\mathcal{G}} r_m\big\{f\big(m(x)\big) - f\big(x\big)\big\} \qquad (x \in S^\Lambda). \tag{1.1}$$

Here $\mathcal{G}$ is a set whose elements are *local maps* $m\colon S^\Lambda \to S^\Lambda$ and $(r_m)_{m\in\mathcal{G}}$ is a collection of nonnegative constants called *rates*, that say with which Poisson intensity the local map m should be applied to the configuration X_t. The precise definitions will be given later in Chapters 2 and 4, but at the moment it suffices to say that if we approximate $(X_t)_{t\geq 0}$ by a discrete-time Markov chain where time is increased in steps of size $\mathrm{d}t$, then

$r_m\,\mathrm{d}t$ is the probability that the map m is applied during the time interval $(t, t + \mathrm{d}t]$.

Often, the lattice Λ has the structure of an (undirected) graph. In this case, we let E denote the corresponding *edge set*. This is a set of unordered pairs $\{i, j\}$ with $i, j \in \Lambda$ and $i \neq j$, that are called *edges*. In drawings of the graph, the fact that $\{i, j\} \in E$ is indicated by connecting the points representing i and j by a line segment. We let

$$\mathcal{E} := \big\{(i, j) : \{i, j\} \in E\big\}$$

denote the corresponding set of all *ordered* pairs (i,j) that correspond to an edge. We call

$$\mathcal{N}_i := \big\{ j \in \Lambda : \{i,j\} \in E \big\} \tag{1.2}$$

the *neighborhood* of the site i.

Many well-known and well-studied interacting particle systems are defined on the *d-dimensional integer lattice* $\mathbb{Z}^d$. We denote the origin by $0 = (0,\dots,0) \in \mathbb{Z}^d$. For any $i = (i_1,\dots,i_d) \in \mathbb{Z}^d$, we let

$$\|i\|_1 := \sum_{k=1}^{d} |i_k| \quad \text{and} \quad \|i\|_\infty := \max_{k=1,\dots,d} |i_k| \qquad (i \in \mathbb{Z}^d)$$

denote the ℓ_1-norm and supremum-norm, respectively. For $R \geq 1$, we set

$$E^d := \big\{\{i,j\} : \|i-j\|_1 = 1\big\} \quad \text{and} \quad E^d_R := \big\{\{i,j\} : 0 < \|i-j\|_\infty \leq R\big\}. \tag{1.3}$$

Then $(\mathbb{Z}^d, E^d)$ is the integer lattice equipped with the *nearest-neighbor* graph structure, and $(\mathbb{Z}^d, E^d_R)$ is the graph obtained by connecting all points within $\|\cdot\|_\infty$-distance R with an edge. We let $\mathcal{E}^d$ and $\mathcal{E}^d_R$ denote the corresponding sets of ordered pairs (i,j).

The graphs we have just introduced have the property that they "look the same" seen from any vertex. An *automorphism* of a graph (Λ, E) is a bijection $\psi\colon \Lambda \to \Lambda$ that "preserves the edges" in the sense that $\{\psi(i), \psi(j)\} \in E$ if and only if $\{i,j\} \in E$. A graph (Λ, E) is called *vertex transitive* if for each pair of vertices $i, j \in \Lambda$, there exists an automorphism ψ such that $\psi(i) = j$. The graphs $(\mathbb{Z}^d, E^d)$ and $(\mathbb{Z}^d, E^d_R)$ we have just introduced are clearly vertex transitive (just take $\psi(k) := k + j - i$). Another example of vertex transitive graphs are d-regular trees, that is, infinite trees in which each vertex has precisely d neighbors.

Before we turn to rigorous mathematical theory, it is good to see a number of examples. It is easy to simulate interacting particle systems on a computer. In simulations, the infinite graphs $(\mathbb{Z}^d, E^d)$ or $(\mathbb{Z}^d, E^d_R)$ are replaced by a finite piece of $\mathbb{Z}^d$, with some choice of the boundary conditions (e.g., periodic boundary conditions).

1.2 The Voter Model

For each $i, j \in \Lambda$, the *voter model map* $\mathrm{vot}_{ij}\colon S^\Lambda \to S^\Lambda$ is defined as

$$\mathtt{vot}_{ij}(x)(k) := \begin{cases} x(i) & \text{if } k = j, \\ x(k) & \text{otherwise.} \end{cases} \tag{1.4}$$

Applying $\mathtt{vot}_{ij}$ to a configuration x has the effect that local state of the site i is copied onto the site j. The *nearest-neighbor voter model* on $\mathbb{Z}^d$ is the interacting particle system with generator

$$G_{\mathrm{vot}} f(x) := \frac{1}{|\mathcal{N}_0|} \sum_{(i,j) \in \mathcal{E}^d} \big\{ f\big(\mathtt{vot}_{ij}(x)\big) - f\big(x\big) \big\} \qquad (x \in S^{\mathbb{Z}^d}). \tag{1.5}$$

Here $\mathcal{N}_0$ is the neighborhood of the origin, and $|\mathcal{N}_0| = 2d$ denotes its cardinality. Similarly, replacing the set of directed edges $\mathcal{E}^d$ by $\mathcal{E}^d_R$ and replacing $\mathcal{N}_0$ by the appropriate set of neighbors in this new graph, we obtain the *range R* voter model.

In the context of the voter model, the local state $x(i)$ at a site i is often called the *type*

Figure 1.1 Four snapshots of a two-dimensional voter model with periodic boundary conditions. Initially, the types of sites are i.i.d. Time evolved in these pictures is 0, 1, 32, and 500.

at i. The voter model is often used to model biological populations, where organisms with different genetic types occupy sites in space. Note that since each site j has $|\mathcal{N}_j| = |\mathcal{N}_0|$ neighbors, the total rate of all maps $\mathtt{vot}_{ij}$ with $i \in \mathcal{N}_j$ is one. In view of this, an alternative way to describe the dynamics in (1.5) is to say that with rate 1, the organism living at a given site dies and is replaced by a descendant chosen with equal probabilities from its neighbors.

An alternative interpretation, that has given the voter model its name, is that sites represent people and types represent political opinions. With rate 1, an individual becomes unsure what political party to vote for, asks a randomly chosen neighbor, and copies his/her opinion.

In Figure 1.1, we see four snapshots of the time evolution of a two-dimensional nearest-neighbor voter model. The initial state is constructed by assigning i.i.d. types to the sites. Due to the copying dynamics, we see patches appear where every site in a local neighborhood has the same type. As time proceeds, these patches, usually called *clusters*, grow in size, so that eventually, for any $N \geq 1$, the probability that all sites within distance N of the origin are of the same type tends to 1.[1]

It turns out that this sort of behavior, called *clustering*, is dimension dependent. The voter model clusters in dimensions 1 and 2, but not in dimensions 3 and more. In Figure 1.2, we see four snapshots of the time evolution of a three-dimensional voter model. The model is

[1] In spite of this, for the model on the infinite lattice, it is still true that the origin changes its type infinitely often.

simulated on a cube with periodic boundary conditions. In this case, we see that even after a long time, there are still many different types near the origin.[2]

Figure 1.2 Four snapshots of a three-dimensional voter model with periodic boundary conditions. Initially, the types of sites are i.i.d. Time evolved in these pictures is 0, 4, 32, and 250.

1.3 The Contact Process

The contact process is another interacting particle system with a biological interpretation. For this process, we choose the local state space $S = \{0, 1\}$. We interpret a site such that $X_t(i) = 1$ as *occupied* by an organism, and a site such that $X_t(i) = 0$ as *empty*. Alternatively, the contact process can be seen as a model for the spread of an infection. In this case, sites with $X_t(i) = 1$ are called *infected* and sites with $X_t(i) = 0$ are called *healthy*.

For each $i, j \in \Lambda$, we define a *branching map* $\mathtt{bra}_{ij} \colon \{0,1\}^\Lambda \to \{0,1\}^\Lambda$ as

$$\mathtt{bra}_{ij}(x)(k) := \begin{cases} x(i) \vee x(j) & \text{if } k = j, \\ x(k) & \text{otherwise.} \end{cases} \tag{1.6}$$

Note that this says that if prior to the application of $\mathtt{bra}_{ij}$, the site i is occupied, then after

[2] On a finite lattice, such as we use in our simulations, one would eventually see one type take over, but the time one has to wait for this to happen is very long compared to dimensions 1 and 2. On the infinite lattice, the probability that the origin has a different type from its right neighbor tends to a positive limit as time tends to infinity.

the application of $\mathtt{bra}_{ij}$, the site j will also be occupied, regardless of its previous state. If initially i is empty, then nothing happens. We interpret this as the organism at i giving *birth* to a new organism at j, or the infected site i *infecting* the site j. If j is already occupied/infected, then nothing happens.

For each $i \in \Lambda$, we also define a *death map* $\mathtt{death}_i : \{0,1\}^\Lambda \to \{0,1\}^\Lambda$ as

$$\mathtt{death}_i(x)(k) := \begin{cases} 0 & \text{if } k = i, \\ x(k) & \text{otherwise.} \end{cases} \tag{1.7}$$

If the map $\mathtt{death}_i$ is applied, then an organism at i, if there is any, dies, respectively, the site i, if it is infected, *recovers* from the infection.

Figure 1.3 Four snapshots of a two-dimensional contact process. Initially, only a single site is infected. The infection rate is 2, the death rate is 1, and time evolved in these pictures is 1, 5, 10, and 20.

Recalling (1.3), the (nearest-neighbor) contact process with *infection rate* $\lambda \geq 0$ and *death rate* $\delta \geq 0$ is the interacting particle system with generator

$$\begin{aligned} G_{\rm cont} f(x) := \lambda & \sum_{(i,j) \in \mathcal{E}^d} \big\{ f\big(\mathtt{bra}_{ij}(x)\big) - f(x) \big\} \\ + \delta & \sum_{i \in \mathbb{Z}^d} \big\{ f\big(\mathtt{death}_i(x)\big) - f(x) \big\} \qquad (x \in \{0,1\}^{\mathbb{Z}^d}). \end{aligned} \tag{1.8}$$

This says that infected sites infect each healthy neighbor with rate λ, and infected sites recover with rate δ.

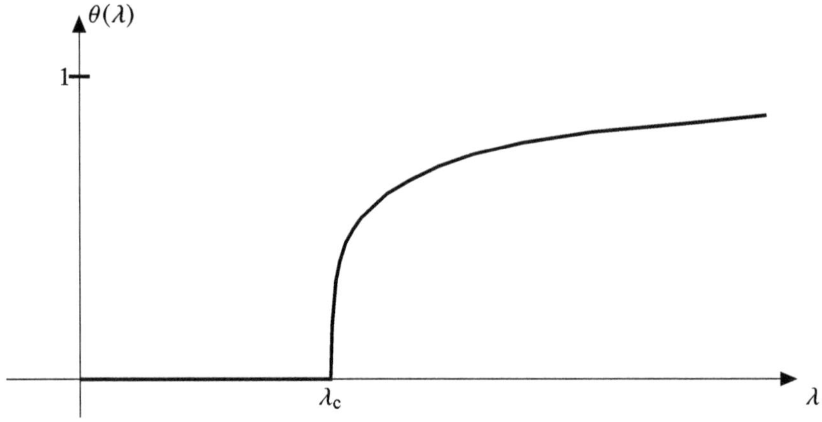

Figure 1.4 Survival probability of the one-dimensional contact process.

In Figure 1.3, we see four snapshots of the time evolution of a two-dimensional contact process. Occupied sites are black and empty sites are white. Initially, only the origin is occupied. The infection rate is 2, and the death rate is 1. In this example, the infection spreads through the whole population, eventually reaching a steady state[3] where a positive fraction of the population is infected. Of course, starting from a single infected site, there is always a positive probability that the infection dies out in the initial stages of the epidemic.

Unlike the voter model, the behavior of the contact process is roughly similar in different dimensions. Instead of the dimension, this time, the proportion λ/δ of the infection rate to the death rate determines the long-time behavior. By varying the speed of time, we can, without loss of generality, choose one of the constants λ and δ to be 1, and it is customary to set $\delta := 1$. Let $e_i \in \{0,1\}^\Lambda$ be defined by $e_i(j) := 1$ if $i = j$ and $:= 0$ otherwise. In Figure 1.4, we have plotted the *survival probability*

$$\theta(\lambda) := \mathbb{P}^{e_0}[X_t \neq 0 \ \forall t \geq 0] \tag{1.9}$$

of the one-dimensional contact process, started in $X_0 = e_0$, that is, with a single infected site at the origin, as a function of the infection rate λ. Note that since $(\mathbb{Z}^d, E^d)$ is vertex transitive, there is nothing special about the origin here: We could have picked any other site instead. For reasons that we cannot explain here, $\theta(\lambda)$ is in fact the same as the probability that the origin is infected in equilibrium; this will be proved in Lemma 6.29.

It turns out that for the nearest-neighbor contact process on $\mathbb{Z}^d$, there exists a *critical value* $\lambda_c = \lambda_c(d)$ with $0 < \lambda_c < \infty$ such that $\theta(\lambda) = 0$ for $\lambda \leq \lambda_c$ and $\theta(\lambda) > 0$ for $\lambda > \lambda_c$. The function θ is continuous, strictly increasing, and concave on $[\lambda_c, \infty)$ and satisfies $\lim_{\lambda\to\infty} \theta(\lambda) = 1$. One has [HHL08, Table 3.2]

$$\lambda_c(1) = 1.648924 \pm 0.00011. \tag{1.10}$$

Proving these statements is not easy, however. For example, continuity of the function θ in

[3] In fact, on the finite square used in our simulations, one can prove that the infection dies out a.s. However, the time one has to wait for this to happen is exponentially large in the system size. For a system of the size shown in Figure 1.3, this time is already too long to be numerically observable.

the point λ_c was proved only in 1990 [BG90], 17 years after the introduction of the model in [CS73, Har74]. The best[4] rigorous upper bound on the constant from (1.10) is $\lambda_c(1) \leq 1.942$, which is proved in [Lig95].

Krone [Kro99] introduced a *two-stage contact process*. In this model, the local state space is $\{0, 1, 2\}$, where 0 represents an empty site, 1 a young organism, and 2 an adult organism. In a branching event, an adult organism produces a young organism on an empty neighboring site. In addition, young organisms can grow up. Both young and adults can die, the young possibly at a higher rate than the adults. The behavior of this model is similar to that of the contact process.

1.4 Ising and Potts Models

In a *stochastic Ising model*, sites in the lattice $\mathbb{Z}^d$ are interpreted as atoms in a crystal, that can have two possible local states, usually denoted by -1 and $+1$. In the traditional interpretation, these states describe the direction of the magnetic field of the atom, and because of this, the local state $x(i)$ of a site i is usually called the *spin* at i. More generally, one can consider *stochastic Potts models* where each "spin" can have $q \geq 2$ possible values. In this case, the local state space is traditionally denoted as $S = \{1, \dots, q\}$, where the special case $q = 2$ corresponds to the Ising model (except for a small difference in notation between $S = \{-1, +1\}$ and $S = \{1, 2\}$).

Given a state x and site i, we let

$$N_{x,i}(\sigma) := \sum_{j \in N_i} 1_{\{x(j) = \sigma\}} \qquad (\sigma \in S) \tag{1.11}$$

denote the number of neighbors of the site i with a spin value $\sigma \in S$. In the Ising and Potts models, sites like or dislike to have the same spin value as their neighbors, depending on a parameter $\beta \in \mathbb{R}$ called the *inverse temperature*. In the physical interpretation of the model, $1/\beta$ corresponds (up to a multiplicative constant) to the temperature (in degrees Kelvin above the absolute zero). Adding a so-called *Glauber dynamics* [Gla63] to the model,[5] sites update their spin values with rate 1, and at such an event, choose a new spin value with probabilities that depend on the values of their neighbors. More precisely, the *stochastic Potts model* with *Glauber dynamics* is the interacting particle system that evolves in such a way that

$$\text{site } i \text{ flips to the value } \sigma \text{ with rate} \quad r_i^\sigma(x) := \frac{e^{\beta N_{x,i}(\sigma)}}{\sum_{\tau \in S} e^{\beta N_{x,i}(\tau)}}. \tag{1.12}$$

More formally, we can write the generator as

$$G_{\text{Potts}} f(x) := \sum_{i \in \mathbb{Z}^d} \sum_{\sigma \in S} r_i^\sigma(x) \big\{ f\big(m_i^\sigma(x)\big) - f\big(x\big) \big\}, \tag{1.13}$$

[4] There exists a sequence of rigorous upper bounds on the constant from (1.10) that is known to converge to the real value, but these bounds are so difficult to calculate that the best bound achieved by this method is much worse than the one in [Lig95].

[5] The terms *Ising model* and *Potts model* refer only to certain Gibbs measures. A *stochastic* Ising model or Potts model is any interacting particle system that has these Gibbs measures as its invariant laws (usually reversible). There exist several different ways to invent a dynamics with this property. This will be explained in a bit more detail in Section 5.6. In this section, we stick to Glauber dynamics.

where $m_i^\sigma : S^\Lambda \to S^\Lambda$ are maps defined by

$$m_i^\sigma(x)(j) := \begin{cases} \sigma & \text{if } j = i, \\ x(j) & \text{otherwise.} \end{cases} \tag{1.14}$$

The attentive reader may notice that the way we have written the generator in (1.13) is different from the way we have written our generators so far, since unlike the rates r_m in (1.1), the rates $r_i^\sigma(x)$ depend on the state x. This will be explained in more detail in Chapter 4. In particular, in Section 4.7, we will see that it is possible to rewrite the generator in (1.13) in a way that fits the general form (1.1) (with rates that do not depend on the state x), but for the Potts model, unlike the models we have seen so far, this way of writing the generator is less natural and more complicated.

Returning to our informal description in (1.12), we notice that for $\beta > 0$, sites prefer to have spin values that agree with as many neighbors as possible, that is, the model is *ferromagnetic*. For $\beta < 0$, the model is *antiferromagnetic*. These terms reflect the situation that in some materials, neighboring spins like to line up, which can lead to long-range order that has the effect that the material can be magnetized. Antiferromagnetic materials, on the other hand, lack this effect.

Alternatively, Potts models can also be interpreted as social or economic models, where sites represent people or firms and spin values represent opinions or the state (financially healthy or not) of a firm [BD01].

In Figure 1.5, we see four snapshots of a two-dimensional nearest-neighbor Potts model with four possible spin values. We have used periodic boundary conditions, and the value of the parameter β is 1.2. Superficially, the behavior is similar to that of a voter model, in the sense that the system forms clusters of growing size that in the end take over any finite neighborhood of the origin. Contrary to the voter model, however, even in the middle of a large cluster that is predominantly of one color, sites can still flip to other values as is clear from (1.12), so in the simulations, we see many small islands of different colors inside large clusters where one color dominates. Another difference is that clustering happens only when the value of the parameter β is large enough. For small values of β, the behavior is roughly similar to the voter model in dimensions $d \geq 3$. There is a critical value $0 < \beta_c < \infty$ where the model changes from one type of behavior to the other. In this respect, the model is similar to the contact process.

To make this critical value visible, imagine that instead of periodic boundary conditions, we would use frozen boundary conditions where the sites at the boundary are kept fixed at one chosen color, say color 1. Then the system has a unique invariant law (equilibrium), in which for sufficiently large values of β, the color 1 is (much) more frequent than the other colors, but for low values of β, all colors occur with (almost) the same frequency. In particular, for the Ising model, where the set of possible spin values is $\{-1, +1\}$, we let

$$m_*(\beta) := \begin{array}{l} \text{the expectation of } x(0) \text{ with } +1 \text{ boundary} \\ \text{conditions, in the limit of large system size.} \end{array} \tag{1.15}$$

This function is called the *spontaneous magnetization*. For the Ising model in two dimensions, the spontaneous magnetization can be explicitly calculated, as was first done by Onsager

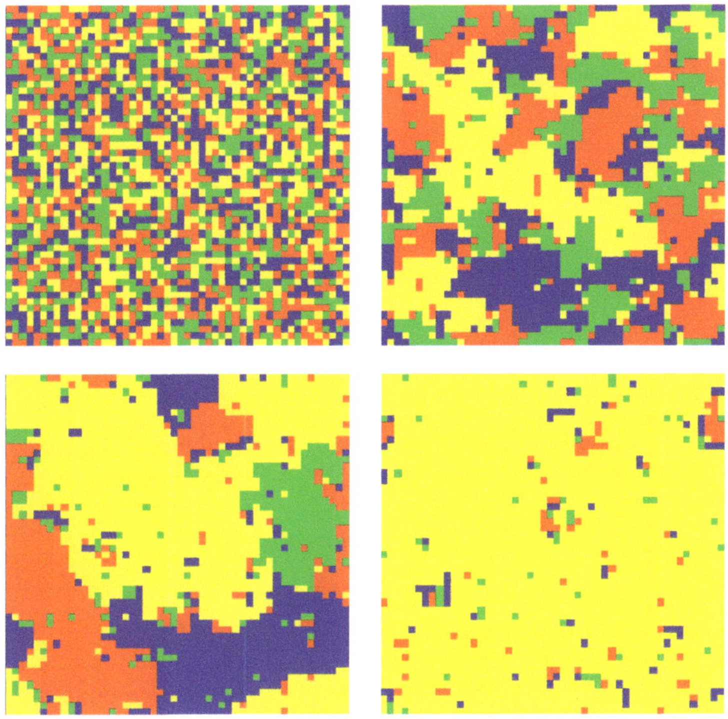

Figure 1.5 Four snapshots of a $q = 4$, $\beta = 1.2$ Potts model with Glauber dynamics and periodic boundary conditions. Initially, the types of sites are i.i.d. Time evolved in these pictures is 0, 4, 32, and 500.

[Ons44]. The formula is

$$m_*(\beta) = \begin{cases} \left(1 - \sinh(\beta)^{-4}\right)^{1/8} & \text{for } \beta \geq \beta_c := \log(1 + \sqrt{2}), \\ 0 & \text{for } \beta \leq \beta_c. \end{cases} \tag{1.16}$$

This function is plotted in Figure 1.6. In this case, the critical point β_c is known explicitly.

For Ising models in dimensions $d \geq 3$, the graph of $m_*(\beta)$ looks roughly similar to Figure 1.6, with $\beta_c \approx 0.442$ in dimension 3 [GPA01], but no explicit formulas are known.

In dimension 1, one has $m^*(\beta) = 0$ for *all* $\beta \geq 0$. More generally, one-dimensional Potts models do not show long-range order, even if β is very large.[6] By this we mean that in equilibrium, the correlation between the spin values at 0 and a point $i \in \mathbb{Z}$ tends to 0 as $i \to \infty$ for any value of β (even though the decay is slow when β is large). In Figure 1.7, we compare the time evolution of a one-dimensional Potts model (with a large value of β) with the time evolution of a one-dimensional voter model. In the voter model, the cluster size keeps growing, but in the Potts model, the typical cluster size converges to a finite limit.

[6] This was first noticed by Ising [Isi25], who introduced the model, but noticed that it was uninteresting, since an incorrect heuristic reasoning led him to believe that what he had proved in dimension 1 would probably hold in any dimension. Peierls [Pei36] realized that dimension matters and proved that the Ising model in higher dimensions does show long-range order.

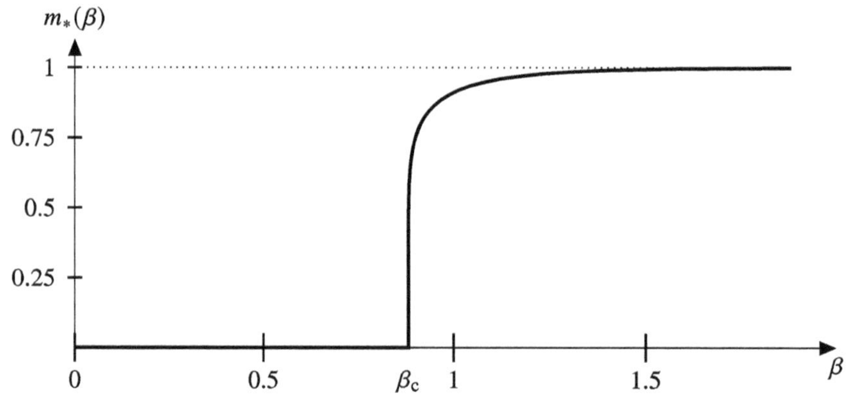

Figure 1.6 The spontaneous magnetization of the two-dimensional Ising model.

1.5 Phase Transitions

Figures 1.4 and 1.6 are examples of a phenomenon that is often observed in interacting particle systems. As a parameter governing the dynamics crosses a particular value, the system goes through an abrupt change in behavior. This is called a *phase transition*, and the value of the parameter is called the *point of the phase transition* or, in the mathematical literature, *critical point*. As we will see in a moment, in the physics literature, the term "critical point" has a more restricted meaning. The term "phase transition" of course also describes the behavior that certain materials change from a gas, fluid, or solid phase into another phase at a particular value of the temperature, pressure, and so forth, and from the theoretical physicist's point of view, this is indeed the same phenomenon.

Both in Figures 1.4 and 1.6, the point of the phase transition in fact separates two regimes, one where the interacting particle systems (on the infinite lattice) have a unique invariant law (below λ_c and β_c) and another regime where there are more invariant laws (above λ_c and β_c). Indeed, for the contact process, the delta measure on the empty configuration is always an invariant law, but above λ_c, a second, nontrivial invariant law also appears. Potts models have q invariant laws (one corresponding to each color) above the critical point.[7] Multiple invariant laws are a general phenomenon associated with phase transitions.

Phase transitions are classified into *first-order* and *second-order* phase transitions.[8] Second-order phase transitions are also called *continuous* phase transitions. The phase transitions in Figures 1.4 and 1.6 are both second order, since the functions θ and m_* are continuous at the critical points λ_c and β_c, respectively. Also, second-order phase transitions are characterized by the fact that at the critical point, there is only one invariant law. By contrast, if we were to draw the function $m_*(\beta)$ of a Potts model for sufficiently large values of q (in dimension

[7] More precisely, they have q invariant laws that have the additional property that they are also translation invariant in space. Depending on the dimension, there may exist additional invariant laws that are not translation invariant.

[8] This terminology was introduced by Paul Ehrenfest. The idea is that in first-order phase transitions, the first derivative of the free energy has a discontinuity, while in a second-order phase transition, the first derivative of the free energy is continuous, and only the second derivative makes a jump.

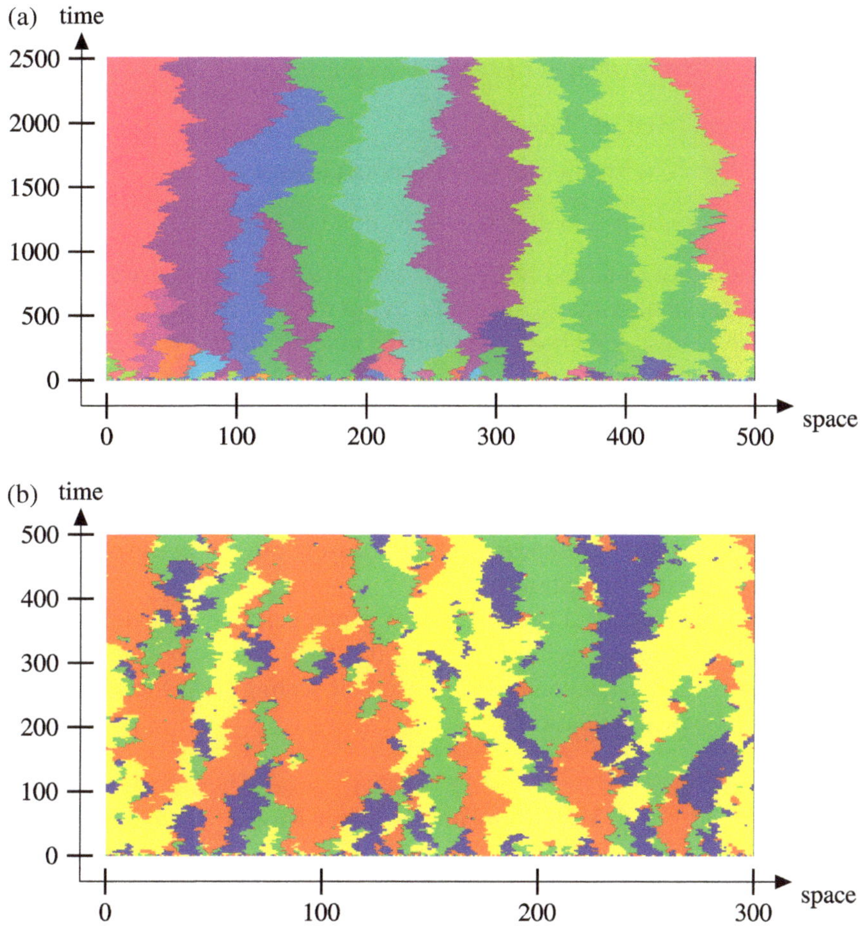

Figure 1.7 Time evolution of a one-dimensional voter model (a) and a one-dimensional Potts model with a high value of β (b).

two, for $q > 4$), then the plot of m_* would make a jump at β_c, and the system would have multiple invariant laws at this point, which means that this phase transition is first order.

It can be difficult to prove that a given phase transition is first or second order. While for the two-dimensional Ising model, continuity of the magnetization follows from Onsager's solution [Ons44], the analogous statement for the three-dimensional Ising model was only proved recently [ADS15] (70 years after Onsager!).

For the Ising model, it is known (but only partially proved) that

$$m_*(\beta) \propto (\beta - \beta_c)^c \quad \text{as } \beta \downarrow \beta_c,$$

where c is a *critical exponent*, which is given by[9]

$$c = 1/8 \text{ in dim } 2, \quad c \approx 0.326 \text{ in dim } 3, \quad \text{and} \quad c = 1/2 \text{ in dim} \geq 4.$$

[9] This exponent is traditionally called β. The values given here are taken from the Wikipedia page *Ising critical exponents* (retrieved 9.2.2025).

For the contact process, it has numerically been observed that

$$\theta(\lambda) \propto (\lambda - \lambda_c)^c \qquad \text{as } \lambda \downarrow \lambda_c,$$

with a critical exponent [HHL08, Table 4.3]

$$\begin{aligned} &c \approx 0.276 \text{ in dim } 1, \quad c \approx 0.583 \text{ in dim } 2, \\ &c \approx 0.813 \text{ in dim } 3, \quad \text{and} \quad c = 1 \text{ in dim} \geq 4. \end{aligned}$$

In theoretical physics, (nonrigorous) *renormalization group theory* is used to explain these critical exponents and calculate them. According to this theory, critical exponents are *universal*. For example, the nearest-neighbor model and the range R models with different values of R all have different values of the critical point, but the critical exponent c has the same value for all these models.[10] Also, changing from the square lattice to, for example, the triangular lattice has no effect on c.

Critical exponents are associated only with second-order phase transitions. At the critical point of a second-order phase transition, one observes *critical behavior*, which involves, for example, power-law decay of correlations. For this reason, physicists use the term "critical point" only for second-order phase transitions.

So far, there is no mathematical theory that can explain critical behavior, except in high dimensions (where one uses a technique called the *lace expansion*) and in a few two-dimensional models (that have a conformally invariant scaling limit that can be described using the Schramm–Loewner equation).

1.6 Variations on the Voter Model

Apart from the models discussed so far, lots of other interacting particle systems have been introduced and studied in the literature to model a plethora of phenomena. Some of these behave very similarly to the models we have already seen (and even appear to have the same critical exponents), while others are completely different. In this and the next sections, we take a brief look at some of these models to get an impression of the possibilities.

The *biased voter model* with *bias* $s \geq 0$ is the interacting particle system with state space $\{0,1\}^{\mathbb{Z}^d}$ and generator (compare (1.5))

$$\begin{aligned} G_{\text{bias}} f(x) := &\frac{1}{|\mathcal{N}_0|} \sum_{(i,j)\in\mathcal{E}^d} \big\{f\big(\texttt{vot}_{ij}(x)\big) - f\big(x\big)\big\} \\ &+ \frac{s}{|\mathcal{N}_0|} \sum_{(i,j)\in\mathcal{E}^d} \big\{f\big(\texttt{bra}_{ij}(x)\big) - f\big(x\big)\big\}, \end{aligned} \tag{1.17}$$

where $\texttt{vot}_{ij}$ and $\texttt{bra}_{ij}$ are the voter and branching maps defined in (1.4) and (1.6). The biased voter model describes a situation where one genetic type of an organism (in this case, type 1) is more fit than the other type, and hence reproduces at a larger rate. Alternatively, this type may represent a new idea or opinion that is more attractive than the current opinion. Contrary to the normal voter model, even if we start with just a single individual of type 1,

[10] Universality in the range R does not always hold. It has been proved that the $q = 3$ ferromagnetic Potts model in dimension two has a first order phase transition for large R [GB07], while the model with $R = 1$ is known to have a second-order phase transition [DST17].

there is a positive probability that type 1 never dies out and indeed takes over the whole population, as can be seen in Figure 1.8.

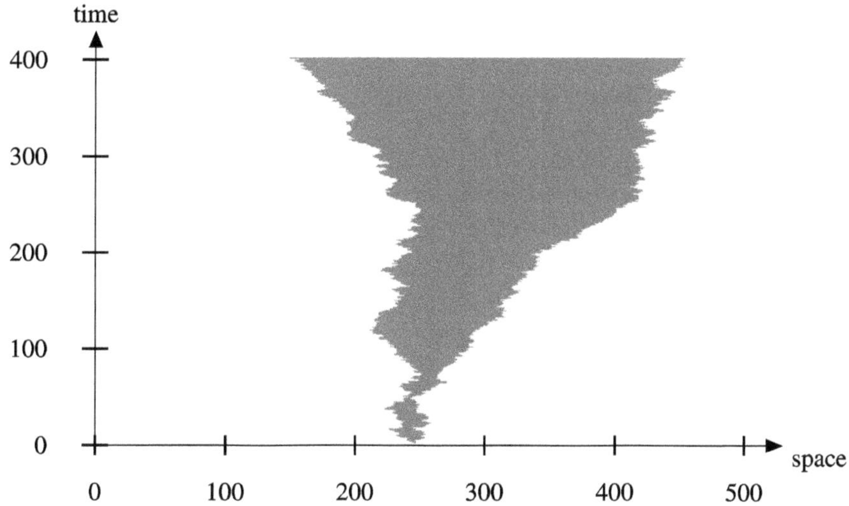

Figure 1.8 Time evolution of a one-dimensional biased voter model with bias $s = 0.2$.

Fix $i \in \mathbb{Z}^d$, and for any $x \in \{0,1\}^{\mathbb{Z}^d}$, let

$$f_\tau(x) := \frac{1}{|\mathcal{N}_i|} \sum_{j \in \mathcal{N}_i} 1_{\{x(j) = \tau\}} \qquad (\tau = 0, 1)$$

be the frequency of type τ in the neighborhood $\mathcal{N}_i$. In the standard voter model, if the present state is x, then the site i changes its type with the following rates:

$$\begin{aligned} 0 &\mapsto 1 \qquad \text{with rate } f_1(x),\\ 1 &\mapsto 0 \qquad \text{with rate } f_0(x). \end{aligned}$$

In the biased voter model, this is changed to

$$\begin{aligned} 0 &\mapsto 1 \qquad \text{with rate } (1+s)f_1(x),\\ 1 &\mapsto 0 \qquad \text{with rate } f_0(x). \end{aligned}$$

Another generalization of the voter model, introduced by Neuhauser and Pacala in [NP99], is defined by the rates

$$\begin{aligned} 0 &\mapsto 1 \qquad \text{with rate } f_1(x)\big(f_0(x) + \alpha f_1(x)\big),\\ 1 &\mapsto 0 \qquad \text{with rate } f_0(x)\big(f_1(x) + \alpha f_0(x)\big), \end{aligned} \tag{1.18}$$

where $0 \le \alpha \le 1$ is a model parameter. Another way of expressing this is to say that if the individual at i is of type τ, then this individual dies with rate

$$f_\tau(x) + \alpha f_{1-\tau}(x), \tag{1.19}$$

and once an individual has died, just as in the normal contact process, it is replaced by a descendant of a uniformly chosen neighbor.

If $\alpha = 1$, then the rate of dying in (1.19) is 1, and we are back at the standard voter model, but for $\alpha < 1$, individuals die less often if they are surrounded by a lot of individuals of the other type. In biology, this models *balancing selection*. This is the effect that individuals who differ from their neighbors experience less competition, which results in a selective drive for high biodiversity.

In the social interpretation of the voter model, we may interpret (1.19) as saying that persons change their mind *less* often if they disagree with a lot of neighbors; that is, the model in (1.18) has "rebellious" behavior.

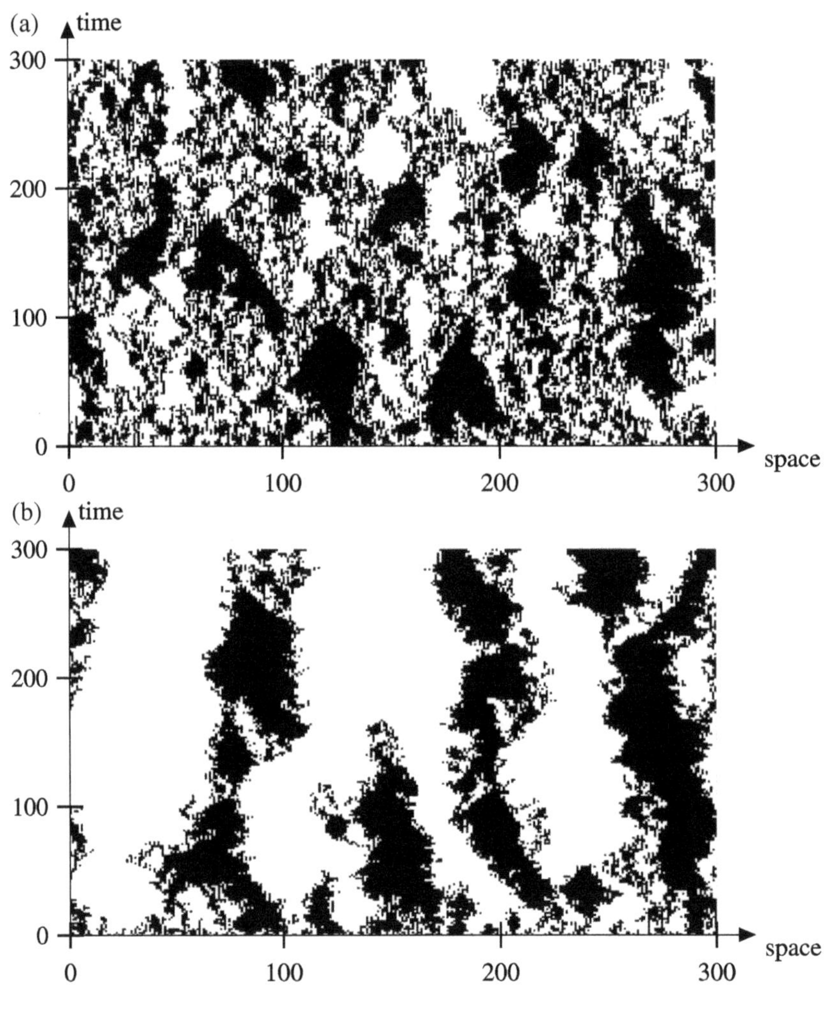

Figure 1.9 Evolution of the Neuhauser–Pacala model with $R = 2$ and $\alpha = 0.2$ (a) and $\alpha = 0.5$ (b).

Numerical simulations, shown in Figure 1.9, suggest that in one dimension, the model in (1.18) with range $R \geq 2$ exhibits a phase transition in α. For α sufficiently close to 1, the model behaves essentially as a voter model, with clusters growing in time, but for small values of α (which represent strong rebellious behavior), the cluster size tends to a finite limit. The latter statement has been rigorously proved, but proving voter-like behavior for α sufficiently close to one is an open problem.

1.7 The Exclusion Process

The exclusion process is a model for traffic or other forms of transport. The local state space is $S = \{0, 1\}$. Sites $i \in \Lambda$ with $x(i) = 1$ are interpreted as being *occupied* by a particle. Sites with $x(i) = 0$ are *empty*. For each $i, j \in \Lambda$ with $i \neq j$, we define an *asymmetric exclusion map* $\texttt{asep}_{ij} \colon S^\Lambda \to S^\Lambda$ by

$$\texttt{asep}_{ij}(x)(k) := \begin{cases} 0 & \text{if } k = i \text{ and } x(j) = 0, \\ x(i) & \text{if } k = j \text{ and } x(j) = 0, \\ x(k) & \text{otherwise.} \end{cases} \tag{1.20}$$

Applying $\texttt{asep}_{ij}$ to a configuration x has the effect that if there is a particle at i and the site j is empty, then the particle at i jumps to j. If there is no particle at i or the site j is already occupied, then nothing happens. Note that these dynamics preserve the number of particles. The one-dimensional lattice $\Lambda = \mathbb{Z}$ is of particular interest. The *asymmetric simple exclusion process* (ASEP) on $\mathbb{Z}$ with parameter $p \in [0, 1]$ is the interacting particle system with generator

$$\begin{aligned} G_{\text{asep}} f(x) := (1-p) \sum_{i \in \mathbb{Z}} \{ f(\texttt{asep}_{i,i-1}) - f(x) \} \\ + p \sum_{i \in \mathbb{Z}} \{ f(\texttt{asep}_{i,i+1}) - f(x) \}. \end{aligned} \tag{1.21}$$

The process with $p = 1$ is called the *totally asymmetric simple exclusion process* (TASEP), and the process with $p = \frac{1}{2}$ is called the *symmetric exclusion process* (SEP), or simply the *exclusion process*. The latter can alternatively also be defined in a different way. For each $i, j \in \Lambda$, we define an *exclusion map* $\texttt{excl}_{ij} \colon S^\Lambda \to S^\Lambda$ by

$$\texttt{excl}_{ij}(x)(k) := \begin{cases} x(j) & \text{if } k = i, \\ x(i) & \text{if } k = j, \\ x(k) & \text{otherwise.} \end{cases} \tag{1.22}$$

Applying $\texttt{excl}_{ij}$ to a configuration x has the effect of interchanging the types of i and j. The interacting particle system with state space $\{0, 1\}^{\mathbb{Z}^d}$ and generator

$$G_{\text{excl}} f(x) = \frac{1}{|\mathcal{N}_0|} \sum_{\{i,j\} \in E^d} \{ f(\texttt{excl}_{ij}(x)) - f(x) \} \qquad (x \in \{0, 1\}^{\mathbb{Z}^d}) \tag{1.23}$$

is called the (symmetric) *exclusion process* on $\mathbb{Z}^d$. One can check that in the one-dimensional case, this is the same process as the one with the generator in (1.21) for $p = \frac{1}{2}$. This follows from the fact that in both processes, the same transitions happen at the same rates. Indeed, from the point of view of how many particles there are on each site, if a particle tries to jump to an already occupied site, then it does not matter if the jump does not take place or the two particles interchange their positions. Mathematically, the equality (in law) of both processes follows from the fact that setting $p = \frac{1}{2}$ in (1.21) and $d = 1$ in (1.23), one has that $G_{\mathrm{asep}} f = G_{\mathrm{excl}} f$ for all functions $f\colon \{0,1\}^{\mathbb{Z}} \to \mathbb{R}$ that depend on finitely many coordinates. This will be proved rigorously in Sections 4.3 and 4.5 of Chapter 4. In the symmetric exclusion process, individual particles move according to random walks that are independent as long as the particles are sufficiently far apart. Particles never meet, and the total number of particles is preserved.

If the totally asymmetric simple exclusion process is started in a deterministic initial state, then its distribution at any later time is a determinantal point process. This means that TASEP is one of the rare examples of an interacting particle system that is, in some sense, explicitly solvable. There are close connections between TASEP, the so-called KPZ universality class (after the Kardar–Parisi–Zhang equation), and random matrix theory. For this reason, TASEP is one of the most studied interacting particle systems, see [Fer13, MQ17].

1.8 Branching and Coalescing Particles

For each $i, j \in \mathbb{Z}^d$, we define a *coalescing random walk map* $\mathtt{rw}_{ij}\colon \{0,1\}^{\mathbb{Z}^d} \to \{0,1\}^{\mathbb{Z}^d}$ by

$$\mathtt{rw}_{ij}(x)(k) := \begin{cases} 0 & \text{if } k = i, \\ x(i) \vee x(j) & \text{if } k = j, \\ x(k) & \text{otherwise.} \end{cases} \tag{1.24}$$

Applying $\mathtt{rw}_{ij}$ to a configuration x has the effect that if the site i is occupied by a particle, then this particle jumps to the site j. If there is already a particle at j, then the two particles coalesce.

The interacting particle system with generator

$$G_{\mathrm{rw}} f(x) = \frac{1}{|\mathcal{N}_0|} \sum_{(i,j)\in\mathcal{E}^d} \{f(\mathtt{rw}_{ij}(x)) - f(x)\} \qquad (x \in \{0,1\}^{\mathbb{Z}^d}) \tag{1.25}$$

describes a system of coalescing random walks, where each particle jumps with rate 1 to a uniformly chosen neighboring site, and two particles on the same site coalesce; see Figure 1.10. Likewise, replacing the coalescing random walk map by the *annihilating random walk map* defined as

$$\mathtt{arw}_{ij}(x)(k) := \begin{cases} 0 & \text{if } k = i, \\ x(i) + x(j) \mod(2) & \text{if } k = j, \\ x(k) & \text{otherwise,} \end{cases} \tag{1.26}$$

yields a system of annihilating random walks that kill each other as soon as two particles land on the same site; see Figure 1.10. If a system of one-dimensional nearest-neighbor coalescing or annihilating random walks is started in a deterministic initial state, then its law at any positive time is a Pfaffian point process [GP+18]. Thus, coalescing or annihilating random walks on $\mathbb{Z}$ are in some sense solvable, similar to TASEP.

The previous two maps (the coalescing and annihilating random walk maps) as well as

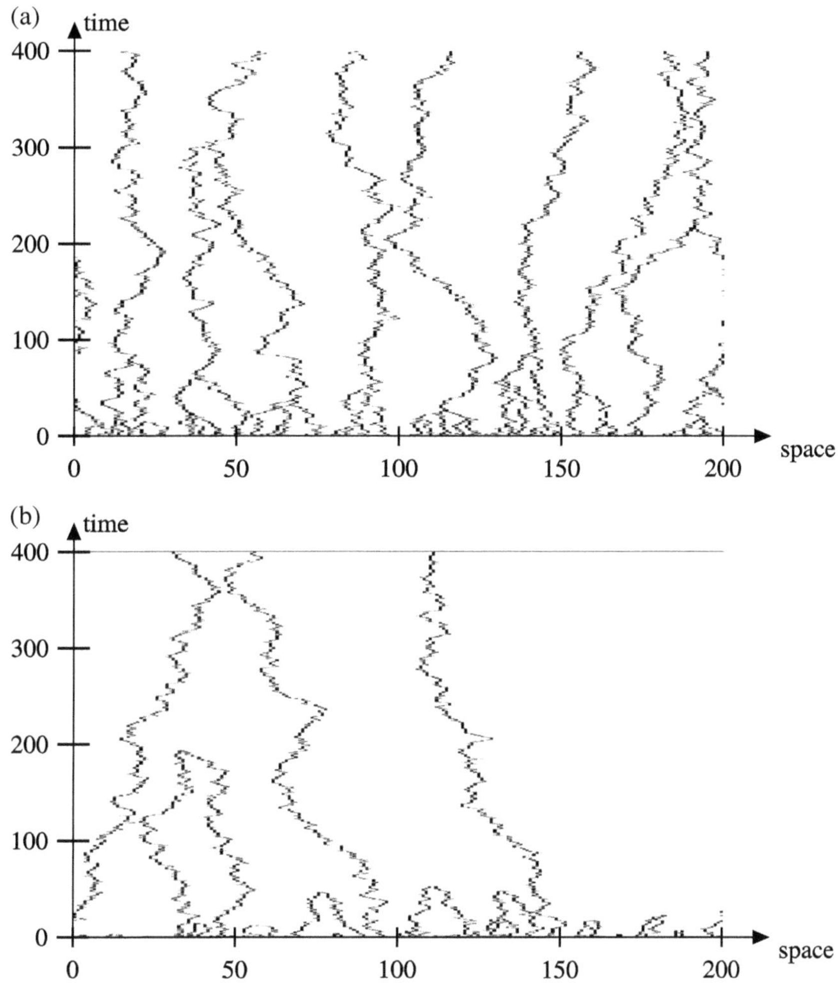

Figure 1.10 Systems of coalescing random walks (a) and annihilating random walks (b).

the exclusion map can be combined with, for example, the branching map and death map from (1.6) and (1.7). In particular, adding coalescing random walk or exclusion dynamics to a contact process models displacement (migration) of organisms. Since in many organisms, you actually need two parents to produce offspring, several authors [Nob92, Dur92, Neu94, SS15] have studied particle systems where the branching map is replaced by the *cooperative branching map*

$$\mathsf{coop}_{ii'j}(x)(k) := \begin{cases} 1 & \text{if } k = j,\ x(i) = 1,\ x(i') = 1, \\ x(k) & \text{otherwise.} \end{cases} \tag{1.27}$$

See Figure 1.11 for a one-dimensional interacting particle system involving cooperative branching and coalescing random walks.

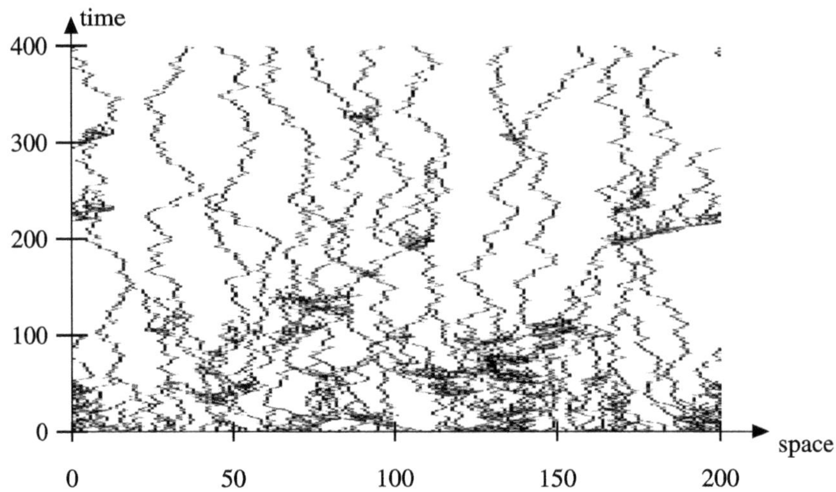

Figure 1.11 A one-dimensional interacting particle system with cooperative branching and coalescing random walk dynamics.

We define a *killing map* by

$$\texttt{kill}_{ij}(x)(k) := \begin{cases} 0 & \text{if } k = j,\ x(i) = 1,\ x(j) = 1, \\ x(k) & \text{otherwise.} \end{cases} \tag{1.28}$$

In words, this says that if there are particles at i and j, then the particle at i kills the particle at j. Sudbury [Sud97, Sud99] has studied a "biased annihilating branching process" with generator of the form

$$\begin{aligned} G_{\text{babp}} f(x) := \lambda &\sum_{(i,j)\in\mathcal{E}^1} \big\{ f\big(\texttt{bra}_{ij}(x)\big) - f\big(x\big) \big\} \\ &+ \sum_{(i,j)\in\mathcal{E}^1} \big\{ f\big(\texttt{kill}_{ij}(x)\big) - f\big(x\big) \big\} \qquad (x \in \{0,1\}^{\mathbb{Z}}). \end{aligned} \tag{1.29}$$

In the physics literature, this model is known as the Fredrickson–Andersen one-spin facilitated model, see formula (26) in [RS03] (with $f = 1$). It is part of the class of *kinetically constrained models* [HT25]. In the mathematical literature on this subject, a slight variant of the model has been studied [BDT19]. Figure 1.12 shows a simulation of such a system when $\lambda = 0.2$. When λ is small, in the simulations, the process seems to behave similarly to systems of branching and coalescing random walks.

1.9 Periodic Behavior

The previous sections served to give a short introduction to some of the main lines of research in interacting particle systems and to showcase how much is known. By contrast, the present section is about a subject that is certainly not, at present, a main line of research and mainly serves to demonstrate how much there still is that we know very little about.

An invariant law of an interacting particle system is a probability distribution on the

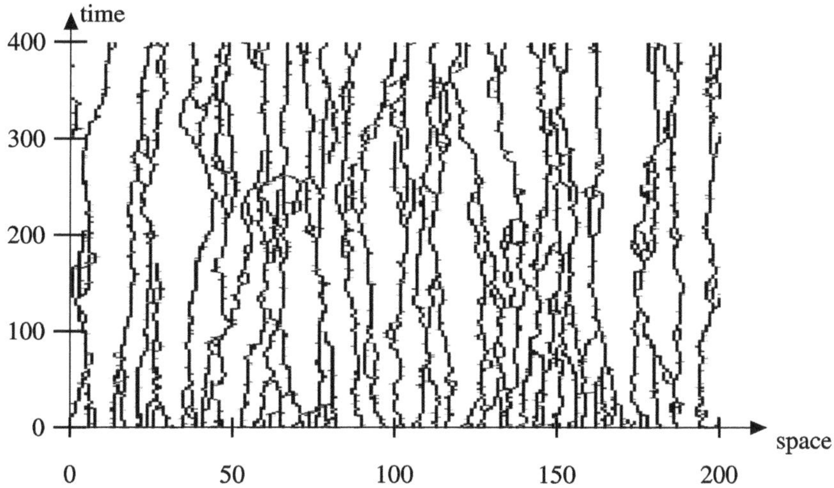

Figure 1.12 A system with branching and killing.

space S^Λ of all possible configurations with the property that if the system at time zero is distributed according to this law, then at all later times it is also distributed according to this law. Invariant laws need not be unique. For example, Potts models above the critical point have q different invariant laws, that are characterized by the color that occupies the majority of the sites.

For all the interacting particle systems and initial states we have considered so far, the system has the property that as time tends to infinity, the distribution of the system converges to an invariant law. This need not always be the case. Perhaps the simplest way in which this can fail is if the system has a *periodic law*, that is, a law that has the property that if the system at time zero is distributed according to this law, then it returns to this law after a finite time $T > 0$ (the *period*), but the system has a different distribution at all intermediate times $0 < t < T$.

Very little is known rigorously about interacting particle systems with periodic laws. Jahnel and Külske [JK14a] have constructed a three-dimensional interacting particle system that has a periodic law. A general result due to Mountford [Mou95] implies that one-dimensional systems with finite-range interactions cannot have periodic laws. Beyond this, almost nothing is known rigorously for spatial models, although there are some studies of periodic behavior in the mean-field limit (see Chapter 3). In particular, it is not known whether periodic laws are possible in two dimensions. The construction in [JK14a] is rather abstract since they do not write down the dynamics of their system explicitly but only prove that such a system exists. Their system also does not have finite range interactions, although the strength of the interaction decays exponentially with distance, which is almost as good.

Numerical simulations suggest that periodic behavior is not a rare phenomenon. Several interacting particle systems with explicit dynamics are known to exhibit periodic behavior in simulations. All known examples seem to work only in dimensions three and higher, however, which suggests that, perhaps, periodic behavior is not possible in two dimensions. This would be in line with (though not rigorously follow from) the Mermin–Wagner theorem from statistical physics that states, informally, that continuous symmetry breaking is not possible in two dimensions [MW66].

Some mechanisms that can lead to periodic behavior are investigated in [DFR13, CFT16, Tov19, CD+20]. A particularly simple model that numerically seems to exhibit periodic behavior is the *cycle conform model* that we describe now. Its local state space is $S = \{0, 1, 2\}$ and it is based on two maps, a *cycle map* and a *conform map*. For each $i \in \Lambda$, the cycle map $\mathtt{cyc}_i : S^\Lambda \to S^\Lambda$ is defined by

$$\mathtt{cyc}_i(x)(k) := \begin{cases} x(i) + 1 \mod(3) & \text{if } k = i, \\ x(k) & \text{otherwise,} \end{cases} \tag{1.30}$$

and for each $i, i', j \in \Lambda$, the conform map $\mathtt{conf}_{ii'j} : S^\Lambda \to S^\Lambda$ is defined by

$$\mathtt{conf}_{ii'j}(x)(k) := \begin{cases} x(i) & \text{if } k = j \text{ and } x(i) = x(i'), \\ x(k) & \text{otherwise.} \end{cases} \tag{1.31}$$

We assume that the lattice Λ has the structure of a graph with set of edges E and that each site has at least two neighbors. For each $j \in \Lambda$, we set (compare Eq. (1.2))

$$\mathcal{N}_j^2 := \{(i, i') : i, i' \in \mathcal{N}_j,\ i \neq i'\}. \tag{1.32}$$

The *cycle conform model* with parameter $\alpha \in [0, 1]$ is the interacting particle system with generator

$$\begin{aligned} G_{\text{cc}} f(x) := (1 - \alpha) \sum_{j \in \Lambda} \{f(\mathtt{cyc}_j(x)) - f(x)\} \\ + \alpha \sum_{j \in \Lambda} \frac{1}{|\mathcal{N}_j^2|} \sum_{(i,i') \in \mathcal{N}_j^2} \{f(\mathtt{conf}_{ii'j}(x)) - f(x)\}. \end{aligned} \tag{1.33}$$

In words, the dynamics can be described as follows. Each site $j \in \Lambda$ becomes active at rate one. With probability $1 - \alpha$, the site j cycles, and with probability α, it conforms. If the site cycles, then it just changes its type to the next type modulo 3. If the site j conforms, then it samples two neighboring sites i and i' at random, and if these happen to have the same type, then the site j copies their type. The idea of this sort of dynamics is to give sites a tendency to conform to the type that is locally in the majority. If we would sample just one neighbor i, then we would obtain voter model dynamics, which gives each type the same chance to spread. By sampling two sites, we introduce a nonlinearity that favors conformation to the local majority.

Numerical simulations on $\mathbb{Z}^d$ in dimensions $d = 1, 2, 3$ suggest the following picture. In dimensions 1 and 2, there is a unique invariant law for each $0 < \alpha < 1$. In dimension 3, there are two critical values $0 < \alpha_c < \alpha'_c < 1$. For $\alpha < \alpha_c$, there is a unique invariant law. For $\alpha > \alpha'_c$, there are three invariant laws that are moreover invariant under translations, in which one of the three local states has a majority. In the intermediate regime $\alpha_c < \alpha < \alpha'_c$, the system exhibits periodic behavior; see Figure 1.13.

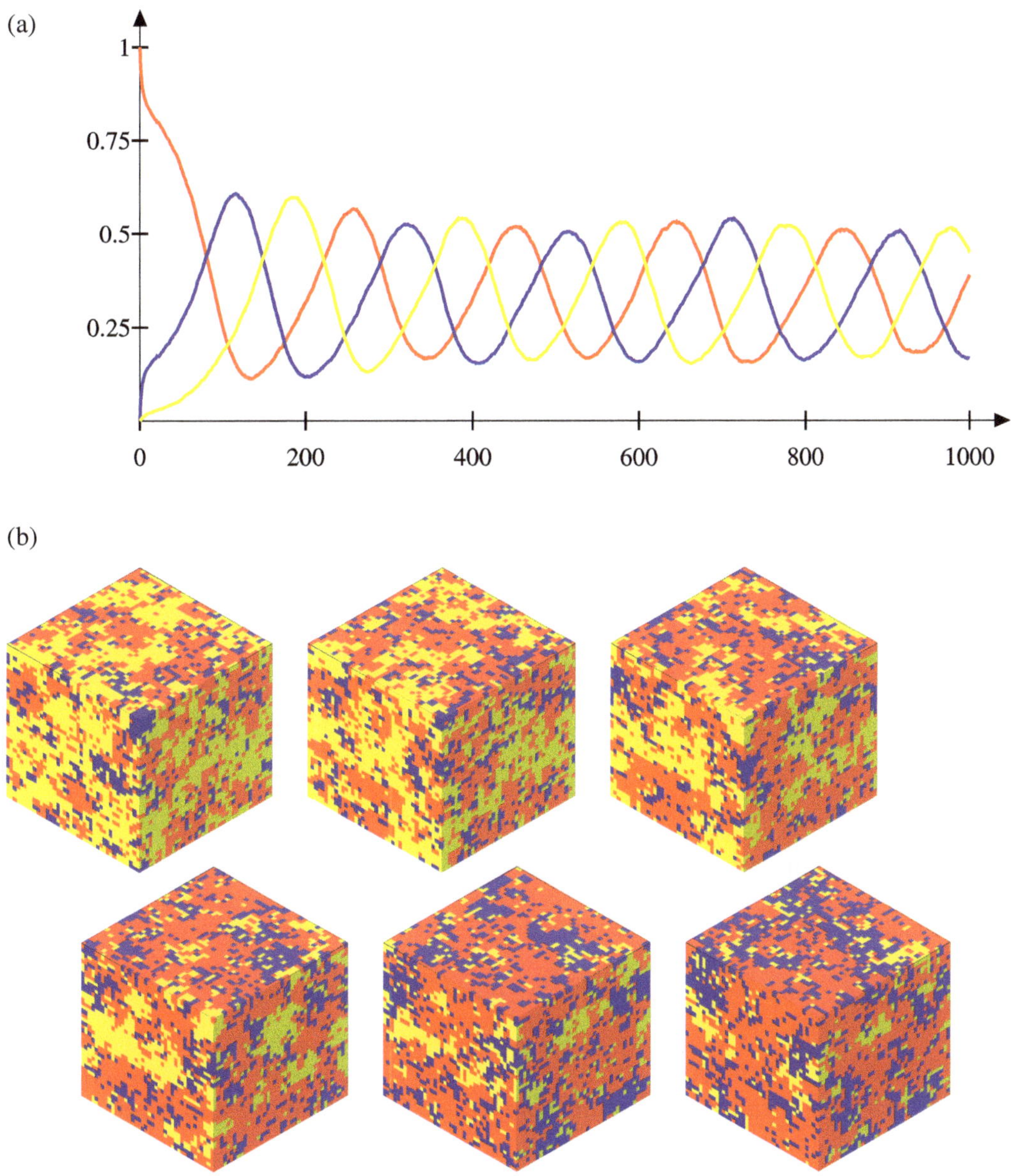

Figure 1.13 Periodic behavior of the cycle conform model with $\alpha = 0.915$ on $\mathbb{Z}^3$ with nearest-neighbor edges. Simulation on a cube with sidelength 40 and periodic boundary conditions. Colors indicate the three states 0, 1, 2. Shown are the frequency of each type as a function of time and the state of the cube at times 601, 613, 626, 638, 651, and 663. Numerically, one sees periodic behavior roughly in the regime $0.903 < \alpha < 0.933$.

2

Continuous-Time Markov Chains

2.1 Finite State Space

Let S be any finite set. A (real) matrix indexed by S is a collection of real constants $A = (A(x, y))_{x,y \in S}$. We calculate with such matrices in the same way as with normal square matrices. Thus, the product AB of two matrices is defined as

$$(AB)(x, z) := \sum_{y \in S} A(x, y)B(y, z) \qquad (x, z \in S).$$

We let 1 denote the identity matrix, which has $1(x, y) = 1$ if $x = y$ and $= 0$ if $x \neq y$. We will sometimes denote this as $1(x, y) = 1_{\{x=y\}}$, where quite generally we let $1_{\{\dots\}}$ denote the indicator of the event $\{\dots\}$, that is, this is one if the conditions inside $\{\dots\}$ are satisfied and zero otherwise. We define the nth power A^n of a matrix indexed by S in the obvious way, with $A^0 := 1$. If $f\colon S \to \mathbb{R}$ is a function, then we also define

$$Af(x) := \sum_{y \in S} A(x, y)f(y) \quad \text{and} \quad fA(y) := \sum_{x \in S} f(x)A(x, y). \tag{2.1}$$

A *matrix semigroup* is a collection of matrices $(A_t)_{t \geq 0}$ such that

$$\lim_{t \downarrow 0} A_t = A_0 = 1 \quad \text{and} \quad A_s A_t = A_{s+t} \qquad (s, t \geq 0).$$

If G is a matrix indexed by S, then setting

$$A_t = e^{tG} := \sum_{n=0}^{\infty} \frac{1}{n!}(tG)^n \qquad (t \geq 0)$$

defines a matrix semigroup $(A_t)_{t \geq 0}$, and conversely every matrix semigroup $(A_t)_{t \geq 0}$ is of this form. The matrix G is called the *generator* of $(A_t)_{t \geq 0}$. The function $t \mapsto A_t$ is continuously differentiable, and one has

$$\tfrac{\partial}{\partial t} A_t = GA_t = A_t G \qquad (t \geq 0). \tag{2.2}$$

A *probability kernel* on S is a matrix $K = (K(x, y))_{x,y \in S}$ such that $K(x, y) \geq 0$ $(x, y \in S)$ and $\sum_{y \in S} K(x, y) = 1$ $(x \in S)$. Clearly, the composition of two probability kernels yields a third probability kernel. A probability kernel is *deterministic* if it is of the form

$$K_m(x, y) := \begin{cases} 1 & \text{if } m(x) = y, \\ 0 & \text{otherwise,} \end{cases}$$

for some function $m\colon S \to S$. It is easy to see that the space of all probability kernels on a finite set S is convex, and the deterministic probability kernels are exactly the extremal elements of this set. It follows that each probability kernel can be written as a convex combination of deterministic probability kernels. Another way to say this is that for each probability kernel K on S, it is possible to find a random map $M\colon S \to S$ such that[1]

$$K(x,y) = \mathbb{P}\big[M(x) = y\big] \qquad (x, y \in S). \tag{2.3}$$

A formula of this form is called a *random mapping representation* of the probability kernel K.

A *Markov semigroup* is a matrix semigroup $(P_t)_{t\geq 0}$, consisting of probability kernels.

Exercise 2.1 (Markov generators) *Let G be a matrix indexed by a finite set S. Show that G generates a Markov semigroup if and only if*

$$G(x,y) \geq 0 \quad (x \neq y) \quad \textit{and} \quad \sum_y G(x,y) = 0. \tag{2.4}$$

By definition, we say that a function f that is defined on an interval $I \subset \mathbb{R}$ is *piecewise constant* if each compact subinterval of I can be divided into finitely many subintervals, so that f is constant on each subinterval. By definition, a *Markov process* with semigroup $(P_t)_{t\geq 0}$ is a stochastic process $X = (X_t)_{t\geq 0}$ with values in S and piecewise constant, right-continuous sample paths, such that

$$\mathbb{P}\big[X_u \in \cdot\,\big|\,(X_s)_{0\leq s\leq t}\big] = P_{u-t}(X_t, \cdot\,) \quad \text{a.s.} \qquad (0 \leq t \leq u). \tag{2.5}$$

Here, on the left-hand side, we condition on the σ-field generated by the random variables $(X_s)_{0\leq s\leq t}$. One can prove that formula (2.5) is equivalent to the statement that

$$\mathbb{P}\big[X_0 = x_0, \ldots, X_{t_n} = x_n\big] = \mathbb{P}[X_0 = x_0] P_{t_1-t_0}(x_0, x_1) \cdots P_{t_n-t_{n-1}}(x_{n-1}, x_n) \tag{2.6}$$

for all $0 = t_0 < t_1 < \cdots < t_n$. From this last formula, we see that for each initial law $\mathbb{P}[X_0 = \cdot\,] = \mu$, there is a unique Markov process with semigroup $(P_t)_{t\geq 0}$ and this initial law. We say that $(P_t)_{t\geq 0}$ are the *transition kernels* of the Markov process. It is custom to let $\mathbb{P}^\mu$ denote the law of the Markov process with initial law μ, and to let $\mathbb{P}^x$ denote the law of the Markov process with deterministic initial state $X_0 = x$ a.s. We let $\mathbb{E}^\mu$ and $\mathbb{E}^x$ denote expectation with respect to $\mathbb{P}^\mu$ and $\mathbb{P}^x$, respectively. Recalling our notation (2.1), we see that for any probability law μ on S and function $f\colon S \to \mathbb{R}$,

$$\begin{aligned} \mu P_t(x) &= \mathbb{P}^\mu[X_t = x], \\ P_t f(x) &= \mathbb{E}^x[f(X_t)] \end{aligned} \qquad (t \geq 0,\ x \in S).$$

In particular, μP_t is the law of the process at time t. We note that

$$\mathbb{P}^x[X_t = y] = P_t(x,y) = 1_{\{x=y\}} + tG(x,y) + O(t^2) \quad \text{as } t \downarrow 0.$$

For $x \neq y$, we call $G(x,y)$ the *rate* of jumps from x to y. Intuitively, if the process is in x, then in the next infinitesimal time interval of length $\mathrm{d}t$ it has a probability $G(x,y)\mathrm{d}t$ to jump to y, independently for all $y \neq x$.

[1] Indeed, this formula says nothing else than $K = \sum_m \mathbb{P}[M = m]K_m$, where the sum runs over all maps $m\colon S \to S$, and K_m is the deterministic kernel defined by the map m.

Let S be a finite set. Let $\mathcal{K}$ be a finite set whose elements are probability kernels on S and let $(r_K)_{K\in\mathcal{K}}$ be nonnegative constants. Then it is straightforward to check that setting

$$Gf := \sum_{K\in\mathcal{K}} r_K \{Kf - f\} \tag{2.7}$$

defines a Markov generator. The following exercise says that conversely, each Markov generator can be written in this form, where we can even choose the set $\mathcal{K}$ so that it has only one element.

Exercise 2.2 *Let S be a finite set. Show that each Markov generator G on S can be written in the form $Gf = r\{Kf - f\}$, where $r \geq 0$ is a constant and K is a probability kernel on S.* Hint: *first add a multiple of the identity matrix to G to make all diagonal entries nonnegative and then normalize.*

If all kernels in the set $\mathcal{K}$ are deterministic, then our expression (2.7) for G takes the form

$$Gf(x) = \sum_{m\in\mathcal{G}} r_m\{f(m(x)) - f(x)\}, \tag{2.8}$$

where $\mathcal{G}$ is a finite set whose elements are maps $m\colon S \to S$ and $(r_m)_{m\in\mathcal{G}}$ are nonnegative constants. This way of writing a generator will come back in formula (2.21) of Section 2.6.

If $(X_t)_{t\geq 0}$ is a Markov process with finite state space S, semigroup $(P_t)_{t\geq 0}$, generator G, and initial law μ, and $f\colon S \to \mathbb{R}$ is a function, then $\mu P_t f = \mathbb{E}^\mu[f(X_t)]$ is the mean of f at time t. But what about the variance of f? It turns out that there is a nice formula for this quantity, that is sometimes useful.[2] For any probability measure μ on S and functions $f, g\colon S \to \mathbb{R}$, we adopt the notation

$$\mathrm{Cov}_\mu(f,g) := \mu(fg) - (\mu f)(\mu g) \quad \text{and} \quad \mathrm{Var}_\mu(f) := \mathrm{Cov}_\mu(f,f). \tag{2.9}$$

In words, $\mathrm{Cov}_\mu(f,g)$ is the covariance of f and g under μ, and $\mathrm{Var}_\mu(f)$ is the variance of f. We define a function $\Gamma_G(f,g)\colon S \to \mathbb{R}$ by[3]

$$\Gamma_G(f,g)(x) := \sum_{y\in S} G(x,y)\{f(y) - f(x)\}\{g(y) - g(x)\}. \tag{2.10}$$

An alternative formula for $\Gamma_G(f,g)$ is

$$\Gamma_G(f,g) = G(fg) - (Gf)g - f(Gg).$$

To see that both formulas are equivalent, we calculate

$$G(fg)(x) = \sum_y G(x,y)f(y)g(y),$$

$$((Gf)g)(x) = \sum_y G(x,y)f(y)g(x), \quad (f(Gg))(x) = \sum_y G(x,y)f(x)g(y).$$

[2] We will actually only need Proposition 2.3 and its infinite-dimensional analogue Proposition 4.35 in Section 5.3, which is not used anywhere else, so the material here can be skipped at a first reading.

[3] This is called the carré du champ operator. Often, a factor $\frac{1}{2}$ is added to its definition because of its relation to Dirichlet forms.

Using the fact that $\sum_y G(x, y) = 0$, this gives

$$\begin{aligned}&\big[G(fg) - (Gf)g - f(Gg)\big](x)\\&\quad = \sum_y G(x, y)\big[f(y)g(y) - f(y)g(x) - f(x)g(y) + f(x)g(x)\big],\end{aligned}$$

which agrees with our first formula for $\Gamma_G(f, g)$. The following formula is well-known (see, e.g., the proofs of [Led00, Thm 2.1] and [Jou07, Lemma 3.3]).

Proposition 2.3 (Covariance formula) *Let $(P_t)_{t\geq 0}$ be the semigroup of a Markov process with finite state space S and generator G. Then, for each probability measure μ on S and functions $f, g\colon S \to \mathbb{R}$, one has*

$$\mathrm{Cov}_{\mu P_t}(f, g) = \mathrm{Cov}_\mu(P_t f, P_t g) + \int_0^t \mathrm{d}s\, \mu P_{t-s}\Gamma_G(P_s f, P_s g).$$

Proof For each $s \in [0, t]$, define $U_s\colon S \to \mathbb{R}$ by

$$U_s := P_{t-s}\big((P_s f)(P_s g)\big).$$

Then

$$\begin{aligned}\mu(U_t - U_0) &= \mu\big((P_t f)(P_t g)\big) - \mu P_t(fg)\\&= \mu\big((P_t f)(P_t g)\big) - (\mu P_t f)(\mu P_t g) + (\mu P_t f)(\mu P_t g) - \mu P_t(fg)\\&= \mathrm{Cov}_\mu(P_t f, P_t g) - \mathrm{Cov}_{\mu P_t}(f, g),\end{aligned}$$

so to complete the proof it suffices to show that

$$U_t - U_0 = \int_0^t \mathrm{d}s\, \tfrac{\partial}{\partial s}U_s = -\int_0^t \mathrm{d}s\, P_{t-s}\Gamma_G(P_s f, P_s g).$$

Using (2.2), we see that

$$\begin{aligned}\tfrac{\partial}{\partial t_1}P_{t_1}\big((P_{t_2}f)(P_{t_3}g)\big) &= P_{t_1}G\big((P_{t_2}f)(P_{t_3}g)\big),\\\tfrac{\partial}{\partial t_2}P_{t_1}\big((P_{t_2}f)(P_{t_3}g)\big) &= P_{t_1}\big((GP_{t_2}f)(P_{t_3}g)\big),\\\tfrac{\partial}{\partial t_3}P_{t_1}\big((P_{t_2}f)(P_{t_3}g)\big) &= P_{t_1}\big((P_{t_2}f)(GP_{t_3}g)\big).\end{aligned}$$

It follows that

$$\tfrac{\partial}{\partial s}U_s = P_{t-s}\big\{\big((GP_s f)(P_s g)\big) + \big((P_s f)(GP_s g)\big) - G\big((P_s f)(P_s g)\big)\big\}$$

which equals $-P_{t-s}\Gamma_G(P_s f, P_s g)$, as required. ∎

Let $X = (X_t)_{t\geq 0}$ be a continuous-time Markov process with finite state space S, generator G, and semigroup $(P_t)_{t\geq 0}$. By definition, an *invariant law* of X is a probability measure ν on S such that

$$\nu P_t = \nu \qquad (t \geq 0).$$

This says that if we start the process in the initial law $\mathbb{P}[X_0 \in \cdot\,] = \nu$, then $\mathbb{P}[X_t \in \cdot\,] = \nu$ for all $t \geq 0$. By definition, the Markov process is *irreducible* if for each $x, y \in S$, there exist $x_0, \ldots, x_n \in S$ with $x = x_0$ and $y = x_n$, such that $G(x_{k-1}, x_k) > 0$ for all $1 \leq k \leq n$. The basic result about invariant laws for continuous-time Markov process with finite state space is the following theorem. A proof can be found in many places, such as, for example, [Lig10, Thm 2.66].

Theorem 2.4 (Convergence to equilibrium) *Let $X = (X_t)_{t\geq 0}$ be a continuous-time Markov process with finite state space S, generator G, and semigroup $(P_t)_{t\geq 0}$. If X is irreducible, then it has a unique invariant law ν. Moreover, one has*

$$\mu P_t \underset{t\to\infty}{\longrightarrow} \nu \quad \textit{for each probability law } \mu \textit{ on } S.$$

If $(P_t)_{t\geq 0}$ is Markov semigroup on a finite set S and ν is an invariant law for $(P_t)_{t\geq 0}$, then it is possible to construct a process $(X_t)_{t\in\mathbb{R}}$ whose finite-dimensional distributions are given by

$$\mathbb{P}\big[X_0 = x_0, \dots, X_{t_n} = x_n\big] = \nu(x_0)P_{t_1-t_0}(x_0, x_1)\cdots P_{t_n-t_{n-1}}(x_{n-1}, x_n) \tag{2.11}$$

for all $t_0 < \cdots < t_n$. Such a process is *stationary*.[4] For this reason, invariant laws are sometimes called *stationary laws*. By definition, an invariant law ν is *reversible* if the stationary process $(X_t)_{t\in\mathbb{R}}$ is equal in law to the time-reversed process $(X_{-t})_{t\in\mathbb{R}}$. It is well-known [Lig10, Exercise 2.44] that this is equivalent to ν satisfying the *detailed balance* equations

$$\nu(x)G(x, y) = \nu(y)G(y, x) \qquad (x, y \in S,\ x \neq y).$$

The left-hand side of this equation can be interpreted as the frequency with which the stationary process jumps from x to y. Detailed balance then says that jumps from x to y happen at the same frequency as jumps from y to x.

An irreducible continuous-time Markov process with finite state space is called *reversible* if its unique invariant law is reversible. Examples are the Ising and Potts models with Glauber dynamics on finite lattices, which have Gibbs measures as their reversible laws. Other examples of processes that have a reversible law are the Fredrickson–Andersen one-spin facilitated model for which a product measure with a suitably chosen intensity is a reversible law, and the symmetric exclusion process for which product measures with arbitrary intensities are reversible laws. Reversibility is a useful property that allows for the use of techniques that are not available for irreversible models, such as Dirichlet form techniques and Poincaré or log-Sobolev inequalities. We refer to [Sal97] as a general introduction to this material and more specifically to [Mar99] for spin systems with Glauber dynamics. Most interacting particle systems we will consider in this book are not reversible.

Exercise 2.5 *Let $X = (X_t)_{t\geq 0}$ be a continuous-time Markov process with finite state space S, generator G, and semigroup $(P_t)_{t\geq 0}$. Let ν be an invariant law and let $(X_t)_{t\in\mathbb{R}}$ be the stationary process from (2.11). Assume that $\nu(x) > 0$ for all $x \in S$. Show that the time-reversed process $(X'_t)_{t\in\mathbb{R}}$ defined as $X'_t := X_{-t}$ is a stationary Markov process and calculate its generator G'.*

[4] Recall that a process $(X_t)_{t\in\mathbb{R}}$ is stationary if for each $s \in \mathbb{R}$, it is equal in distribution to $(X'_t)_{t\in\mathbb{R}}$ defined as $X'_t := X_{t-s}$ $(t \in \mathbb{R})$.

2.2 The Embedded Markov Chain

Continuous-time Markov processes with countable state space (also known as continuous-time Markov chains) can in many ways be treated in the same way as those with a finite state space, but there are some complications. The first complication one has to deal with is that such processes may explode. The second complication is that their long-time behavior is more complicated than in the finite case: They may fail to have invariant laws, or have invariant (in particular reversible) measures that are infinite. For this reason, they need to be distinguished into positive recurrent, null recurrent, and transient processes. A good general reference for this material is [Lig10, Chapter 2].

We calculate with matrices indexed by a countably infinite set S in the same way as for finite S, provided the infinite sums are well-defined (i.e., not of the form $\infty - \infty$). Generalizing our earlier definition, we say that K is a *subprobability kernel* if $\sum_y K(x,y) \leq 1$ for all $x \in S$. Also when S is infinite, we define generators as in (2.4). Note that

$$\sum_{y \in S} G(x,y) = G(x,x) + \sum_{y:\, y \neq x} G(x,y).$$

Since $G(x,y) \geq 0$ for $x \neq y$, the infinite sum on the right-hand side is always well-defined, though a priori it may be infinite. The condition $\sum_y G(x,y) = 0$ says that it must be finite and equal to $-G(x,x)$, however. In the special context of continuous-time Markov chains, a generator is traditionally called a *Q-matrix* (and denoted as Q) but we will stick to the term generator.

It is well known [Lig10, Section 2.5.2] that one can construct a continuous-time Markov chain with generator G from its associated embedded discrete Markov chain and independent, exponentially distributed holding times. We now recall this construction. Let G be a generator, let $c(x) := -G(x,x)$ $(x \in S)$, and let K be the probability kernel on S defined by

$$K(x,y) := \begin{cases} c(x)^{-1} G(x,y) & \text{if } c(x) > 0,\ x \neq y, \\ 1 & \text{if } c(x) = 0,\ x = y, \\ 0 & \text{otherwise.} \end{cases}$$

For each $x \in S$, let $(Y^x_k)_{k \geq 0}$ be the discrete-time Markov chain with initial state $Y^x_0 = x$ and transition kernel K. Set $N := \inf\{n \geq 0 : c(Y^x_n) = 0\}$, which may be infinite, and let $(\sigma_k)_{k \geq 0}$ be i.i.d. exponentially distributed random variables with mean one, independent of $(Y^x_k)_{k \geq 0}$. We define $(\tau_k)_{0 \leq k \leq N+1}$ by

$$\tau_0 := 0, \quad \tau_n := \sum_{k=0}^{n-1} \sigma_k / c(Y^x_k) \quad (1 \leq n \leq N+1),$$

where we use the conventions that $\sigma_k / c(Y^x_k) := \infty$ if $c(Y^x_k) = 0$ and $N + 1 := \infty$ if $N = \infty$. We set $\tau := \tau_{N+1}$. Note that $\tau = \infty$ on the event that $N < \infty$, but τ may be finite on the event that $N = \infty$. We define a stochastic process $(X^x_t)_{t \geq 0}$ with values in $S_\infty := S \cup \{\infty\}$ by

$$X^x_t := \begin{cases} Y^x_k & \text{if } t \in [\tau_k, \tau_{k+1}),\ 0 \leq k < N+1, \\ \infty & \text{if } t \geq \tau. \end{cases}$$

We call τ the *explosion time*. We set $X^\infty_t := \infty$ $(t \geq 0)$ and define probability kernels $(\overline{P}_t)_{t \geq 0}$

on S_∞ by

$$\overline{P}_t(x,y) := \mathbb{P}\big[X^x_t = y\big] \qquad (t \geq 0,\ x, y \in S_\infty).$$

We call $(X^x_t)_{t\geq 0}$ the *continuous-time Markov chain* with generator G, and we call $(Y^x_k)_{k\geq 0}$ its associated *embedded discrete-time Markov chain*. It is well-known [Lig10, Section 2.5.2] that $(X^x_t)_{t\geq 0}$ is a Markov process (in the sense of (2.5) and (2.6)) with state space S_∞ and transition kernels $(\overline{P}_t)_{t\geq 0}$. In Exercise 2.6, you will be asked to prove this. An alternative proof will be suggested in Exercise 2.23. The random times

$$\eta_k := \sigma_k / c(Y^x_k) \quad (0 \leq k < N)$$

are called the *holding times*. Note that conditional on the embedded chain $(Y^x_k)_{k\geq 0}$, the holding times $(\eta_k)_{0\leq k<N}$ are independent exponentially distributed such that η_k has mean $1/c(Y^x_k)$. If $\tau = \infty$ a.s. for each initial state $x \in S$, then we say that the continuous-time Markov chain with generator G is *nonexplosive*. In the opposite case, it is *explosive*. We let

$$P_t(x,y) := \overline{P}_t(x,y) \qquad (t \geq 0,\ x, y \in S) \tag{2.12}$$

denote the restrictions of the transition kernels $(\overline{P}_t)_{t\geq 0}$ to S. If G is explosive, then these are only subprobability kernels.

Exercise 2.6 *Show that $(X^x_t)_{t\geq 0}$ is a Markov process in the sense of (2.5) with state space S_∞ and transition kernels $(\overline{P}_t)_{t\geq 0}$.* Hint: *let M be the number of jumps of the process $(X^x_s)_{0\leq s\leq t}$, which may be infinite. Then, after conditioning on $(X^x_s)_{0\leq s\leq t}$, you know M as well as $(Y^x_k)_{0\leq k<M+1}$ and $(\sigma_k)_{0\leq k<M}$, plus in the case that $M < \infty$ you have the information that $\sum_{k=0}^{M} \sigma_k / c(Y^x_k) > t$. Given all this information, what do you know about the process $(X^x_u)_{u\geq t}$?*

2.3 Generator Construction

Let G be the generator of a continuous-time Markov chain with countable state space S. Generalizing our earlier definition to countable state spaces, we call the collection of subprobability kernels $(P_t)_{t\geq 0}$ defined in (2.12) the *Markov semigroup* with generator G. In this section, we make a more direct link between $(P_t)_{t\geq 0}$ and G.

We calculate with infinite matrices as in the finite case. We observe that if $f\colon S \to \mathbb{R}$ is nonnegative, then all terms in the infinite sum $Gf(x) := \sum_{y\in S} G(x,y) f(y)$ except one are nonnegative so

$$Gf\colon S \to (-\infty, \infty] \text{ is well-defined for all } f\colon S \to [0,\infty). \tag{2.13}$$

Let $u\colon S \times [0,\infty) \to [0,\infty)$ be a function. We say that u solves the *Kolmogorov backward equation*

$$\tfrac{\partial}{\partial t} u_t(x) = \sum_y G(x,y) u_t(y) \qquad (t \geq 0,\ x \in S), \tag{2.14}$$

if the function $t \mapsto u_t(x)$ is continuously differentiable for each $x \in S$ and (2.14) holds. By (2.13), the right-hand side of (2.14) is well-defined, and the equality in (2.14) implies that it must be finite. We say that u is a *minimal* solution to (2.14) if any other solution u' with the same initial condition $u'_0 = u_0$ satisfies $u_t(x) \leq u'_t(x)$ for all $t \geq 0$ and $x \in S$. Note that for a given initial condition, there can be at most one minimal solution to (2.14). The following theorem is the main result of this section.

Theorem 2.7 (Generator construction) *Let G be the generator of a continuous-time Markov chain with countable state space S and let $(P_t)_{t\geq 0}$ be the Markov semigroup with generator G. Then, for each bounded function $f: S \to [0,\infty)$, the function*

$$u_t(x) := P_t f(x) \qquad (t \geq 0,\ x \in S)$$

is the minimal solution to the Kolmogorov backward equation (2.14) with initial condition f.

To prepare for the proof of Theorem 2.7, as a first step, we set $c(x) := -G(x,x)$ $(x \in S)$, and we consider the equation

$$u_t(x) = u_0(x)e^{-c(x)t} + \int_0^t \mathrm{d}s\, e^{-c(x)s} \sum_{y:\, y\neq x} G(x,y)u_{t-s}(y). \tag{2.15}$$

By definition, a solution to (2.15) is a function $u: S\times[0,\infty) \to [0,\infty)$ such that $t \mapsto u_t(x)$ is measurable for all $x \in S$ and (2.15) holds for all $t \geq 0$ and $x \in S$. Note that since $G(x,y) \geq 0$ for $x \neq y$, the sum over y and consequently also the integral over s are well-defined, even though a priori the outcome may be ∞ (a posteriori, of course, (2.15) implies that the outcome must be finite).

Lemma 2.8 (First jump decomposition) *Under the assumptions of Theorem 2.7, the function u solves (2.15) with $u_0 = f$.*

Proof Since f is bounded $u_t(x) := P_t f(x) < \infty$ for all $t \geq 0$ and $x \in S$. (This is the only place in the proof where we use the boundedness of f.) Let $(X_t^x)_{t\geq 0}$ be the continuous-time Markov chain with generator G and initial state $x \in S$, constructed from the embedded discrete-time Markov chain $(Y_k^x)_{k\geq 0}$ and i.i.d. standard exponential random variables $(\sigma_k)_{k\geq 0}$ as in the previous section. We extend f to S_∞ by setting $f(\infty) := 0$ so that $P_t f(x) = \mathbb{E}[f(X_t^x)]$. Let τ_k denote the time when $(X_t^x)_{t\geq 0}$ makes its k-th jump. If $c(x) = 0$, then $\tau_1 = \infty$ a.s. and $X_t = x$ a.s. for all $t \geq 0$, which implies $u_t(x) = P_t f(x) = \mathbb{E}[f(X_t)] = f(x)$ $(t \geq 0)$, so (2.15) is trivially satisfied with $u_0(x) = f(x)$. We assume, therefore, from now on that $c(x) > 0$. In this case, conditional on $\tau_1 = s$ and $X_{\tau_1}^x = y$, the process $(X_{\tau_1+t}^x)_{t\geq 0}$ is equally distributed with $(X_t^y)_{t\geq 0}$, which allows us to write

$$\begin{aligned}
\mathbb{E}[f(X_t^x)] &= f(x)\mathbb{P}[\tau_1 > t] + \int_0^t \mathbb{P}[\tau_1 \in \mathrm{d}s] \sum_{y:\, y\neq x} \mathbb{P}[X_{\tau_1}^x = y]\mathbb{E}\big[f(X_{t-s}^y)\big] \\
&= f(x)e^{-c(x)t} + \int_0^t c(x)e^{-c(x)s}\mathrm{d}s\, c(x)^{-1} \sum_{y:\, y\neq x} G(x,y)P_{t-s}f(y),
\end{aligned}$$

which shows that $u_t := P_t f$ solves (2.15) with $u_0 = f$. ∎

Lemma 2.9 (The backward equation) *If a function $u: S \times [0,\infty) \to [0,\infty)$ solves (2.14), then it solves (2.15). Conversely, each bounded solution to (2.15) also solves (2.14).*

Proof If u solves (2.14), then

$$\tfrac{\partial}{\partial t}u_t(x) + c(x)u_t(x) = \sum_{y:\, y\neq x} G(x,y)u_t(y),$$

which implies

$$\tfrac{\partial}{\partial t}\Big(e^{c(x)t}u_t(x)\Big) = e^{c(x)t}\sum_{y:\,y\neq x} G(x,y)u_t(y) \qquad (t \geq 0,\ x \in S).$$

Integrating and then multiplying both sides of the equation by $e^{-c(x)t}$, we obtain

$$u_t(x) = u_0(x)e^{-c(x)t} + \int_0^t \mathrm{d}s\, e^{-c(x)(t-s)} \sum_{y:\,y\neq x} G(x,y)u_s(y),$$

which after the substitution $s \mapsto t - s$ yields (2.15).

Conversely, if (2.15) holds and u is bounded, then the right-hand side is continuous in t for each x and hence so is the left-hand side. But then the right-hand side must actually be continuously differentiable as a function of t, and the same must be true for the left-hand side. We can then reverse the argument above (differentiating instead of integrating) to obtain (2.14). ∎

Lemma 2.10 (Comparison principle) *Let G be the generator of a continuous-time Markov chain with countable state space S and let $(P_t)_{t\geq 0}$ be the Markov semigroup with generator G. Assume that $u\colon S \times [0,\infty) \to [0,\infty)$ satisfies*

$$\tfrac{\partial}{\partial t}u_t(x) \geq \sum_{y} G(x,y)u_t(y) \qquad (t \geq 0,\ x \in S),$$

where $t \mapsto u_t(x)$ is continuously differentiable for each $x \in S$. Then

$$P_t u_0(x) \leq u_t(x) \qquad (t \geq 0,\ x \in S).$$

Proof Let $(X_t^x)_{t\geq 0}$ be the continuous-time Markov chain with generator G and initial state x and as in Section 2.2, let τ_k denote the time of its kth jump, for $1 \leq k \leq N$, where N denotes the total number of jumps, which may be finite or infinite, and let τ denote the explosion time. Then

$$P_t u_0(x) = \mathbb{E}\big[u_0(X_t^x)1_{\{t<\tau\}}\big] \qquad (t \geq 0).$$

On the event that $N < \infty$ we set $\tau_k := \infty$ for $k > N$, and we define

$$u_0^{(n)}(x) = \mathbb{E}\big[u_0(X_t^x)1_{\{t<\tau_n\}}\big] \qquad (t \geq 0).$$

Then

$$u_t^{(n)}(x) \underset{n\to\infty}{\longrightarrow} P_t u_0(x) \qquad (t \geq 0).$$

Using the same argument as in the proof of Lemma 2.8, we see that

$$u_t^{(n+1)}(x) = u_0(x)e^{-c(x)t} + \int_0^t \mathrm{d}s\, e^{-c(x)s} \sum_{y:\,y\neq x} G(x,y)u_{t-s}^{(n)}(y).$$

By the same argument as in the proof of Lemma 2.9, with all equalities replaced by inequalities,

$$u_t(x) \geq u_0(x)e^{-c(x)t} + \int_0^t \mathrm{d}s\, e^{-c(x)s} \sum_{y:\,y\neq x} G(x,y)u_{t-s}(y).$$

We claim that $u_t^{(n)}(x) \leq u_t(x)$ $(n \geq 0,\ t \geq 0)$. The proof is by induction. Clearly $u_t^{(0)}(x) = 0 \leq u_t(x)$ $(t \geq 0)$. Assuming that the statement holds for n, we have

$$\begin{aligned}u_t(x) &\geq u_0(x)e^{-c(x)t} + \int_0^t \mathrm{d}s\, e^{-c(x)s} \sum_{y:\,y\neq x} G(x,y)u_{t-s}(y)\\ &\geq u_0(x)e^{-c(x)t} + \int_0^t \mathrm{d}s\, e^{-c(x)s} \sum_{y:\,y\neq x} G(x,y)u_{t-s}^{(n)}(y) = u_t^{(n+1)}(x).\end{aligned}$$

Letting $n \to \infty$, we obtain $u_t(x) \geq P_t(x)$. ∎

Proof of Theorem 2.7 By Lemmas 2.8 and 2.9, u solves the Kolmogorov backward equation (2.14) with initial condition f. If u' is another solution, then Lemma 2.10 implies that $u \leq u'$, showing that u is minimal. ∎

We conclude this section with the following observation.

Proposition 2.11 (Uniqueness of solutions) *The Kolmogorov backward equation (2.14) has a unique bounded solution u with initial condition $u_0 = f$ for each bounded function $f\colon S \to [0,\infty)$ if and only if the continuous-time Markov chain with generator G is nonexplosive.*

Proof For each $r \in \mathbb{R}$, let $\underline{r}\colon S \to \mathbb{R}$ denote the function that is constantly equal to r. If G is explosive, then $u_t := P_t\underline{1}$ and $u'_t := \underline{1}$ $(t \geq 0)$ are two different bounded solutions of the Kolmogorov backward equation (2.14) with initial condition $u_0 = u'_0 = \underline{1}$, proving that solutions are not unique.

On the other hand, assume that G is nonexplosive, and that u is a bounded solution with initial condition $u_0 = f$. Since u is bounded, there exists an $r \geq 0$ such that $u_t \leq \underline{r}$ $(t \geq 0)$. Lemma 2.10 tells us that $P_t f \leq u_t$ $(t \geq 0)$. Also, since $\underline{r} - u$ solves (2.14) with initial condition $\underline{r} - f$, Lemma 2.10 tells us that $\underline{r} - u_t \geq P_t(\underline{r} - f) = \underline{r} - P_t f$ $(t \geq 0)$ where in the last step, we have used that G is nonexplosive. Combining these inequalities, we see that $u_t = P_t f$ $(t \geq 0)$. ∎

2.4 Lyapunov Functions

It is tempting to think of explosive continuous-time Markov chains as pathological, but there exist very natural chains that are explosive. In fact, each transient chain can with a suitable random time transformation be transformed into an explosive chain, so from this point of view, the distinction between transient and recurrent chains would appear to be more fundamental than the distinction between explosive and nonexplosive chains. Nevertheless, it is useful to have at our disposal a technique for proving that a given chain is nonexplosive. In the present section, we will show how nonexplosiveness can be proved with the help of Lyapunov functions. Below is the main result of this section. The term "Lyapunov function" originates in the stability theory of ordinary differential equations but is sometimes also used for certain functions occurring in Foster's theorem, that gives necessary and sufficient conditions for positive recurrence of a Markov chain. The role of the function L in the following theorem is similar, so using the term in a general sense, we may call it a Lyapunov function too. Note that GL in condition (ii) below is well-defined by (2.13).

Theorem 2.12 (Sufficient conditions for nonexplosiveness) *Let G be the generator of a continuous-time Markov chain with countable state space S and let $c(x) := -G(x,x)$ $(x \in S)$. Assume that there exists a function $L: S \to [0,\infty)$ and constant $\lambda \in \mathbb{R}$ such that:*

(i) $\sup\{c(x) : x \in S,\ L(x) < C\}$ *is finite for all* $C < \infty$,
(ii) $GL \leq \lambda L$.

Then the continuous-time Markov chain $(X_t)_{t\geq 0}$ with generator G is nonexplosive and

$$\mathbb{E}^x\big[L(X_t)\big] \leq e^{\lambda t} L(x) \qquad (t \geq 0,\ x \in S).$$

The proof of Theorem 2.12 depends on two lemmas.

Lemma 2.13 (Exponential bound) *Let $(X_t)_{t\geq 0}$ be a continuous-time Markov chain with generator G, started in $X_0 = x$. Assume that $L: S \to [0,\infty)$ satisfies $GL \leq \lambda L$ for some $\lambda \in \mathbb{R}$. Then*

$$\mathbb{E}^x\big[L(X_t) 1_{\{t<\tau\}}\big] \leq e^{\lambda t} L(x) \qquad (t \geq 0), \tag{2.16}$$

where τ denotes the explosion time of $(X_t)_{t\geq 0}$.

Proof The function $u_t(x) := L(x)e^{\lambda t}$ satisfies $\frac{\partial}{\partial t} u_t \geq G u_t$ $(t \geq 0)$, so Lemma 2.10 tells us that $P_t u_0 \leq u_t$ $(t \geq 0)$, which is the same as (2.16). ∎

Lemma 2.14 (Bounded jump rates) *Let G be the generator of a continuous-time Markov chain with countable state space S and let $c(x) := -G(x,x)$ $(x \in S)$. Assume that $\sup_{x\in S} c(x) < \infty$. Then G is nonexplosive.*

Proof Let $(X_t^x)_{t\geq 0}$ be the continuous-time Markov chain with generator G and initial state x, constructed from its embedded discrete-time Markov chain $(Y_k^x)_{k\geq 0}$ and i.i.d. standard exponential random variables $(\sigma_k)_{k\geq 0}$ as in Section 2.2. Let N be the total number of jumps, which may be finite or infinite, and let τ denote the explosion time. On the event that $N < \infty$, we have $\tau = \infty$, while on the event that $N = \infty$, we have

$$\tau = \sum_{k=0}^{\infty} \sigma_k / c(Y_k^x).$$

By our assumption that $C := \sup_{x\in S} c(x) < \infty$, we can estimate this from below by

$$\tau \geq C^{-1} \sum_{k=1}^{\infty} \sigma_k$$

which is ∞ a.s. by the strong law of large numbers. ∎

Proof of Theorem 2.12 It suffices to prove that G is nonexplosive, since the statement about the expectation of $L(X_t)$ then follows from Lemma 2.13. We set

$$S_C := \big\{x \in S : L(x) < C\big\}$$

and define a generator G_C by

$$G_C(x,y) := \begin{cases} G(x,y) & \text{if } x \in S_C, \\ 0 & \text{if } x \notin S_C \end{cases}$$

We let $(X_t^x)_{t\geq 0}$ and $(X_t^{x,C})_{t\geq 0}$ denote the continuous-time Markov chains with generators G and G_C, respectively. It follows from the construction of these processes in terms of their embedded Markov chains that we can naturally couple these processes such that

$$X_t^x = X_t^{x,C} \qquad \forall t \leq \tau_C := \inf\{t \geq 0 : X_t^x \notin S_C\}.$$

In fact, we then have

$$X_t^{x,C} = X_{t\wedge\tau_C}^x \qquad (t \geq 0),$$

that is, $(X_t^{x,C})_{t\geq 0}$ corresponds to the process $(X_t^x)_{t\geq 0}$ stopped as soon as it leaves S_C.

Let τ denote the explosion time of $(X_t^x)_{t\geq 0}$. Lemma 2.14 and assumption (i) of the theorem imply that $(X_t^{x,C})_{t\geq 0}$ is nonexplosive. Since the processes $(X_t^x)_{t\geq 0}$ and $(X_t^{x,C})_{t\geq 0}$ are equal up to time τ_C it follows that $\tau_C \leq \tau$. Making λ larger if necessary, we can without loss of generality assume that $\lambda \geq 0$. Assumption (ii) of the theorem then implies that also $G_C L \leq \lambda L$. Indeed, $G_C L(x) = GL(x)$ if $L(x) < C$ and $G_C L(x) = 0$ otherwise. We can therefore use Lemma 2.13 and the fact that G_C is nonexplosive to conclude that

$$C\mathbb{P}^x[\tau_C \leq t] \leq \mathbb{E}\big[L(X_t^{x,C})\big] \leq e^{\lambda t} L(x) \qquad (t \geq 0).$$

Since $\tau_C \leq \tau$, it follows that

$$\mathbb{P}^x[\tau \leq t] \leq e^{\lambda t} L(x)/C \qquad (t \geq 0),$$

so letting $C \to \infty$ we see that $\tau = \infty$ a.s. ∎

2.5 Poisson Point Sets

The construction of a continuous-time Markov chain from its embedded discrete-time Markov chain is useful for theoretical purposes, but for the purpose of studying interacting particle systems, a different construction, that is based on Poisson point sets, will turn out to be much more useful. To prepare for this, in the present section, we recall the definition of Poisson point sets and some of their basic properties.

Let S be a σ-compact[5] metrizable space. We will mainly be interested in the case that $S = \mathcal{G} \times \mathbb{R}$ where $\mathcal{G}$ is a countable set. We let $\mathcal{S}$ denote the Borel-σ-field on S. A *locally finite measure* on $(S, \mathcal{S})$ is a measure μ such that $\mu(C) < \infty$ for all compact $C \subset S$.

Let $(\Omega, \mathcal{F}, \mathbb{P})$ be our underlying probability space. A random measure on S is a function $\xi \colon \Omega \times \mathcal{S} \to [0, \infty]$ such that for fixed $\omega \in \Omega$, the function $\xi(\omega, \cdot\,)$ is a locally finite measure on $(S, \mathcal{S})$, and for fixed $A \in \mathcal{S}$, the function $\xi(\,\cdot\,, A)$ is measurable. By [Kal97, Lemma 1.37], we can think of ξ as a random variable with values in the space of locally finite measures on $(S, \mathcal{S})$, equipped with the σ-field generated by the maps $\mu \mapsto \mu(A)$ with $A \in \mathcal{S}$. Then the integral $\int f \mathrm{d}\xi$ defines a $[0, \infty]$-valued random variable for all measurable $f \colon S \to [0, \infty]$. There exists a unique measure, denoted by $\mathbb{E}[\xi]$, such that

$$\int f \,\mathrm{d}\mathbb{E}[\xi] = \mathbb{E}\Big[\int f \,\mathrm{d}\xi\Big]$$

for all measurable $f \colon S \to [0, \infty]$. The measure $\mathbb{E}[\xi]$ is called the *intensity* of ξ.

[5] This means that there exists a countable collection of compact sets $S_i \subset S$ such that $\bigcup_i S_i = S$.

The following result follows from [Kal97, Lemma 10.1 and Prop. 10.4].[6] Below, $\mathcal{S}_{\text{loc}} := \{A \in \mathcal{S} : \overline{A} \text{ is compact}\}$ denotes the set of measurable subsets of S whose closure is compact.

Proposition 2.15 (Poisson point measures) *Let μ be a locally finite measure on $(S, \mathcal{S})$. Then there exists a random measure ξ, unique in distribution, such that for any disjoint $A_1, \ldots, A_n \in \mathcal{S}_{\text{loc}}$, the random variables $\xi(A_1), \ldots, \xi(A_n)$ are independent and $\xi(A_i)$ is Poisson distributed with mean $\mu(A_i)$.*

We call a random measure ξ satisfying the conditions of Proposition 2.15, a *Poisson point measure* with *intensity* μ. Indeed, one can check that $\mathbb{E}[\xi] = \mu$. We note that $\xi(A) \in \mathbb{N}$ for all $A \in \mathcal{S}_{\text{loc}}$. Such measures are called (locally finite) *counting measures*. Each locally finite counting measure ν on S is of the form

$$\nu = \sum_{x \in \text{supp}(\nu)} n_x \delta_x,$$

where $\text{supp}(\nu)$, the support of ν, is a locally finite subset of S, the n_x are positive integers, and δ_x denotes the delta-measure at x. We say that ν is *simple* if $n_x = 1$ for all $x \in \text{supp}(\nu)$. Recall that a measure μ has an *atom* at x if $\mu(\{x\}) > 0$. A measure μ is called *atomless* if it has no atoms, that is, $\mu(\{x\}) = 0$ for all $x \in S$. The already mentioned [Kal97, Prop. 10.4] tells us the following.

Lemma 2.16 (Simple Poisson point measures) *Let ξ be a Poisson point measure with locally finite intensity μ. Then ξ is a.s. simple if and only if μ is atomless.*

If μ is atomless, then a Poisson point measure ξ with intensity μ is characterized by its support $\omega := \text{supp}(\xi)$. We call ω a *Poisson point set* with intensity μ. Intuitively, ω is a set such that $\mathbb{P}[\omega \cap \mathrm{d}x \neq \emptyset] = \mu(\mathrm{d}x)$, independently for each infinitesimal subset $\mathrm{d}x \subset S$.

For any counting measure ν on S and measurable function $f : S \to [0, 1]$, we introduce the notation

$$f^\nu := \prod_{i \in I} f(x_i) \qquad \text{with} \quad \nu = \sum_{i \in I} \delta_{x_i},$$

where the index set I is either finite or countably infinite and, by definition, a product of zero factors is 1. Thus $f^0 := 1$, where 0 denotes the counting measure that is identically zero. Alternatively, our definition says that

$$f^\nu = e^{\int (\log f) \mathrm{d}\nu},$$

where $\log 0 := -\infty$ and $e^{-\infty} := 0$. It is easy to see that $f^\nu f^{\nu'} = f^{\nu + \nu'}$.

Lemma 2.17 (Laplace functionals) *Let μ be a locally finite measure on $(S, \mathcal{S})$ and let ξ be a Poisson point measure with intensity μ. Then*

$$\mathbb{E}\big[(1 - f)^\xi\big] = e^{-\int f \mathrm{d}\mu} \tag{2.17}$$

for each measurable $f : S \to [0, 1]$. Conversely, if ξ is a random counting measure and (2.17) holds for all continuous, compactly supported f, then ξ is a Poisson point measure with intensity μ.

[6] In fact, [Kal97, Prop. 10.4] shows that it is possible to construct Poisson point measures on arbitrary measurable spaces, assuming only that the intensity measure is σ-finite, but we will not need this generality.

Proof The fact that Poisson point measures satisfy (2.17) is proved in [Kal97, Lemma 10.2], which is written in terms of $-\log f$, rather than f. The fact that knowing (2.17) for all continuous, compactly supported f determines the law of ξ uniquely follows from [Kal97, Lemma 10.1]. ∎

Formula (2.17) can be interpreted in terms of thinning. Consider a counting measure $\nu = \sum_i \delta_{x_i}$, let $f: S \to [0, 1]$ be measurable, and let χ_i be independent *Bernoulli random variables* (i.e., random variables with values in $\{0, 1\}$) with $\mathbb{P}[\chi_i = 1] = f(x_i)$. Then the random counting measure

$$\nu' := \sum_i \chi_i \delta_{x_i}$$

is called an *f-thinning* of the counting measure ν. Note that

$$\mathbb{P}[\nu' = 0] = \prod_i \mathbb{P}[\chi_i = 0] = (1 - f)^\nu.$$

In view of this, the left-hand side of (2.17) can be interpreted as the probability that after thinning the random counting measure ξ with f, no points remain.

We cite the following fact from [Kal97, Lemma 10.17].

Lemma 2.18 (Poisson points on the half-line) *Let $0 < c < \infty$ and let ℓ denote the Lebesgue measure on $[0, \infty)$. Let $(\tau_k)_{k\geq 0}$ be real random variables such that $\tau_0 = 0$ and $\sigma_k := \tau_k - \tau_{k-1} > 0$ $(k \geq 1)$. Then $\omega := \{\tau_k : k \geq 1\}$ is a Poisson point set on $[0, \infty)$ with intensity $c\ell$ if and only if the random variables $(\sigma_k)_{k\geq 1}$ are i.i.d. exponentially distributed with mean c^{-1}.*

We will need the following property of Poisson sets.

Proposition 2.19 (Markov property of Poisson sets) *Let S be a countable set, let μ be a locally finite measure on S, and let ω be a Poisson point set on $S \times [0, \infty)$ with intensity measure*

$$\rho\big(\{x\} \times [s, t]\big) := \mu(\{x\})(t - s) \qquad (x \in S,\ 0 \leq s \leq t).$$

Let $S' \subset S$ and assume that $0 < \mu(S') < \infty$. Set

$$\tau := \inf\{\tau_x : x \in S'\} \quad \textit{with} \quad \tau_x := \inf\big\{t \geq 0 : (x, t) \in \omega\big\} \qquad (x \in S').$$

Then a.s., there exists a unique $X \in S'$ such that $\tau_X = \tau$. Setting

$$\omega' := \big\{(x, t - \tau) : (x, t) \in \omega,\ t > \tau\big\},$$

one has that the random variables τ, X, and ω' are independent,

$$\mathbb{P}[\tau \geq t] = e^{-\mu(S')t} \quad (t \geq 0), \qquad \mathbb{P}[X = x] = \frac{\mu(\{x\})}{\mu(S')} \quad (x \in S'), \tag{2.18}$$

and ω' is equally distributed with ω.

Proof We first prove the statement if $S' = S$. Let $\lambda := \mu(S)$ and $\pi := \mu/\lambda$. Let $(\sigma_k)_{k\geq 1}$

be i.i.d. exponentially distributed random variables with parameter λ and let $\tau_k := \sum_{i=1}^{k} \sigma_i$ $(k \geq 1)$. Let $(X_k)_{k\geq 1}$ be i.i.d. with law π and independent of $(\sigma_k)_{k\geq 1}$. Set

$$\xi := \sum_{k=1}^{\infty} \delta_{\tau_k} \quad \text{and} \quad \eta := \sum_{k=1}^{\infty} \delta_{(X_k,\tau_k)}.$$

By Lemma 2.18, ξ is a Poisson point measure on $[0,\infty)$ with intensity measure $\lambda\ell$. We claim that η is a Poisson point measure on $S\times[0,\infty)$ with intensity measure ρ as in the proposition. To see this, we apply Lemma 2.17. Let $f\colon S\times\mathbb{R} \to [0,1]$ be continuous and compactly supported. Define $\overline{f}\colon [0,\infty) \to [0,1]$ by

$$\overline{f}(t) := \sum_{x\in S} \pi(x) f(x,t) \qquad (t \geq 0). \tag{2.19}$$

Then, using the fact that ξ is a Poisson point measure on $[0,\infty)$ with intensity measure $\lambda\ell$, we see that

$$\begin{aligned}\mathbb{E}\big[(1-f)^{\eta}\big] &= \mathbb{E}\Big[\prod_{k=1}^{\infty}\big(1-f(X_k,\tau_k)\big)\Big] = \prod_{k=1}^{\infty}\mathbb{E}\big[1-f(X_k,\tau_k)\big]\\ &= \prod_{k=1}^{\infty}\mathbb{E}\big[1-\overline{f}(\tau_k)\big] = \mathbb{E}\big[(1-\overline{f})^{\xi}\big] = e^{-\lambda\int_0^{\infty}\overline{f}(t)\,\mathrm{d}t} = e^{-\int f\,\mathrm{d}\rho}.\end{aligned}$$

It follows that

$$\omega := \big\{(X_k,\tau_k) : k \geq 1\big\}$$

is a Poisson point set on $S\times[0,\infty)$ with intensity measure ρ. Now clearly $\tau = \tau_1$ and $X = X_1$ are distributed as in (2.18), while

$$\omega' = \big\{(X_k,\tau_k-\tau_1) : k \geq 2\big\}$$

is independent of (X_1,τ_1), and equally distributed with ω. This completes the proof in the special case that $S' = S$. The general case follows immediately by applying what we have already proved to the restriction of ω to $S'\times[0,\infty)$ and then using that this is independent of the restriction of ω to $(S\backslash S')\times[0,\infty)$. ∎

Exercise 2.20 *Let ξ_1,ξ_2 be independent Poisson point measure with intensities μ_1,μ_2. Show that $\xi_1+\xi_2$ is a Poisson point measures with intensity $\mu_1+\mu_2$.* Hint: *Lemma 2.17.*

2.6 Poisson Construction of Markov Processes

In the present section, we will show how a continuous-time Markov chain with countable state space S can be constructed by applying certain maps $m\colon S\to S$ at the times of a Poisson point process. We start with the following observation.

Lemma 2.21 (Random mapping representation) *Let S be a countable set, let $\mathcal{G}$ be a countable collection of maps $m\colon S\to S$, and let $(r_m)_{m\in\mathcal{G}}$ be nonnegative real numbers such that*

$$c(x) := \sum_{m\colon m(x)\neq x} r_m < \infty \quad \textit{for all } x\in S. \tag{2.20}$$

Then there exists a generator G of a continuous-time Markov chain with state space S such that

$$Gf(x) = \sum_{m\in\mathcal{G}} r_m\{f(m(x)) - f(x)\} \tag{2.21}$$

for all bounded $f: S \to [0, \infty)$. Conversely, each generator G of a continuous-time Markov chain with state space S can be written in the form (2.21) for a suitable choice of a collection $\mathcal{G}$ of maps $m: S \to S$ and nonnegative real numbers $(r_m)_{m\in\mathcal{G}}$ satisfying (2.20).

Proof Let $\mathcal{G}$ be a countable collection of maps $m: S \to S$, let $(r_m)_{m\in\mathcal{G}}$ be nonnegative real numbers satisfying (2.20), and let $c(x)$ be as defined in (2.20). Then it is straightforward to check that

$$G(x, y) := \begin{cases} \displaystyle\sum_{\substack{m\in\mathcal{G}\\ m(x)=y}} r_m & \text{if } x \neq y, \\ -c(x) & \text{if } x = y \end{cases} \qquad (x, y \in S)$$

defines a generator such that (2.21) holds. To see that each generator G can be written in this form, we define for each $x, y \in S$ with $x \neq y$ a map $m_{x,y}: S \to S$ by

$$m_{x,y}(z) := \begin{cases} y & \text{if } z = x, \\ z & \text{otherwise,} \end{cases}$$

we set $\mathcal{G} := \{m_{x,y} : x, y \in S,\ x \neq y\}$ and $r_{m_{x,y}} := G(x, y)$. It is then straightforward to check that the rates $(r_m)_{m\in\mathcal{G}}$ satisfy (2.20) and (2.21) holds. ∎

We call the way of writing G as in (2.21) a *random mapping representation* of G. Recall that random mapping representations of probability kernels have been defined in Section 2.1. There is a close connection between the two. Indeed, we will see that if $(P_t)_{t\geq 0}$ is the Markov semigroup with generator G, then using a random mapping representation of G we can for each $t \geq 0$ construct a random mapping representation of P_t. We will do this by giving a Poisson construction of the continuous-time Markov chain with generator G. In the context of interacting particle systems, such Poisson constructions are known as *graphical representations*, and they have many applications.

From now on, we assume that G is the generator of a continuous-time Markov chain with countable state space S and we fix a random mapping representation of G of the form (2.21) in terms of rates $(r_m)_{m\in\mathcal{G}}$ satisfying (2.20). We equip the space $\mathcal{G} \times \mathbb{R}$ with the measure

$$\rho\big(\{m\} \times [s, t]\big) := r_m(t - s) \qquad \big(m \in \mathcal{G},\ s \leq t\big). \tag{2.22}$$

Let ω be a Poisson point set with intensity ρ. We call ω the *graphical representation* associated with the random mapping representation (2.21). We claim that for each $t \in \mathbb{R}$, there exists at most one $m \in \mathcal{G}$ such that $(m, t) \in \omega$. To see this, we note that for each $m \in \mathcal{G}$, the set

$$\xi_m := \sum_{t:\,(m,t)\in\omega} \delta_t$$

is a Poisson point measure on $\mathbb{R}$ with intensity $r_m\ell$, where ℓ denotes the Lebesgue measure. Since the sets $\mathbb{R} \times \{m\}$ $(m \in \mathcal{G})$ are disjoint, the random measures ξ_m $(m \in \mathcal{G})$ are independent, and hence by Exercise 2.20, for each $m \neq m'$, the measure $\xi_m + \xi_{m'}$ is a Poisson point measure on $\mathbb{R}$ with intensity $(r_m + r_{m'})\ell$. Since the Lebesgue measure is atomless, by Lemma 2.16, this Poisson point measure is simple, so there are no times $t \in \mathbb{R}$ for which both $(m,t) \in \omega$ and $(m',t) \in \omega$. In view of this, we can unambiguously define a random function $\mathbb{R} \ni t \mapsto \mathfrak{m}_t^\omega \in \mathcal{G}$ by setting

$$\mathfrak{m}_t^\omega := \begin{cases} m & \text{if } (m,t) \in \omega, \\ 1 & \text{otherwise,} \end{cases}$$

where we write 1 to denote the identity map.

As before, we write $S_\infty := S \cup \{\infty\}$. We equip S_∞ with a topology such that a set $A \subset S_\infty$ is closed if and only if A is either finite or A is infinite and $\infty \in A$. One can check that S_∞ is compact in this topology and a sequence $x_n \in S$ converges to ∞ if and only if it leaves every finite subset of S, that is, for each finite $S' \subset S$, there exists an m such that $x_n \notin S'$ for all $n \geq m$. The topological space S_∞ is known as the *one-point compactification* of S. We extend the maps $m \in \mathcal{G}$ to S_∞ by putting $m(\infty) := \infty$ $(m \in \mathcal{G})$.

Fix $s \in \mathbb{R}$. By definition, we say that a random function $X \colon [s,\infty) \to S_\infty$ solves the evolution equation

$$X_t = \mathfrak{m}_t^\omega(X_{t-}) \qquad (t > s) \tag{2.23}$$

if

(i) $\lim_{r \downarrow t} X_r = X_t$ $(t \geq s)$ and $\lim_{r \uparrow t} X_r =: X_{t-}$ exists $(t > s)$,
(ii) $X_t = \infty$ for all $t \geq \tau := \inf\{r \geq s : X_r = \infty\}$,

and (2.23) holds. Below is the main result of this section.

Theorem 2.22 (Stochastic flow) *Let G be the generator of a continuous-time Markov chain with countable state space S and let ω be the graphical representation associated with a random mapping representation of G. Then almost surely, for each $s \in \mathbb{R}$ and $x \in S_\infty$, there exists a unique solution $(X_t^{s,x})_{t \geq s}$ to the evolution equation (2.23) with initial state $X_s^{s,x} = x$. Setting*

$$\mathbf{X}_{s,t}(x) := X_t^{s,x} \qquad (s \leq t,\ x \in S_\infty) \tag{2.24}$$

defines a collection of maps $(\mathbf{X}_{s,t})_{s \leq t}$ from S_∞ into itself such that

$$\mathbf{X}_{s,s} = 1 \quad \text{and} \quad \mathbf{X}_{t,u} \circ \mathbf{X}_{s,t} = \mathbf{X}_{s,u} \qquad (s \leq t \leq u). \tag{2.25}$$

If G is nonexplosive, then almost surely, $\mathbf{X}_{s,t}$ maps S into itself for all $s \leq t$. If $s \in \mathbb{R}$ and X_0 is an S-valued random variable with law μ, independent of ω, then the process $(X_t)_{t \geq 0}$ defined as

$$X_t := \mathbf{X}_{s,s+t}(X_0) \qquad (t \geq 0) \tag{2.26}$$

is distributed as the continuous-time Markov chain with generator G and initial law μ.

Formula (2.25) says that the random maps $(\mathbf{X}_{s,t})_{s \leq t}$ form a *stochastic flow*. This stochastic flow is *stationary* in the sense that

$$\mathbf{X}_{0,t} \text{ is equally distributed with } \mathbf{X}_{s,s+t} \quad (s \in \mathbb{R}).$$

We note that since $\mathbf{X}_{s,t}$ is constructed using only Poisson points of the form (m, r) with $s < r \leq t$, and restrictions of a Poisson point set to disjoint parts of the space are independent, it follows that the stochastic flow $(\mathbf{X}_{s,t})_{s \leq t}$ has *independent increments* in the sense that

$$\mathbf{X}_{t_0,t_1}, \ldots, \mathbf{X}_{t_{n-1},t_n} \quad \text{are independent} \quad \forall\, t_0 < \cdots < t_n.$$

We note that (2.26) implies that

$$P_t(x, y) = \mathbb{P}\big[\mathbf{X}_{s,s+t}(x) = y\big] \qquad (s, t \in \mathbb{R},\ x, y \in S),$$

so indeed, as announced, we have found a random mapping representation of the subprobability kernels $(P_t)_{t \geq 0}$.

Proof of Theorem 2.22 We start by proving that almost surely, for each $s \in \mathbb{R}$ and $x \in S_\infty$, there exists a unique solution $(X_t^{s,x})_{t \geq s}$ to the evolution equation (2.23) with initial state $X_s^{s,x} = x$. If $x = \infty$, then clearly $X_t^\infty = \infty$ $(t \geq s)$ is the unique solution of (2.23) so without loss of generality, we assume from now on that $x \in S$. For each $x \in S$, the set

$$\big\{t \in \mathbb{R} : \mathfrak{m}_t^\omega(x) \neq x\big\}$$

is a Poisson point set on $\mathbb{R}$ with intensity $c(x)$ defined in (2.20), which is finite by assumption. This allows us to inductively define times $(\tau_n)_{n \geq 0}$ and a discrete chain $(Y_n^x)_{n \geq 0}$ by setting $\tau_0 := 0$, $Y_0^x := x$,

$$\tau_{n+1} := \begin{cases} \inf\big\{t > \tau_n : \mathfrak{m}_t^\omega(x_n) \neq x_n\big\} & \text{if } \tau_n < \infty, \\ \infty & \text{if } \tau_n = \infty, \end{cases}$$

and

$$Y_{n+1}^x := \begin{cases} \mathfrak{m}_{\tau_{n+1}}^\omega(Y_n^x) & \text{if } \tau_{n+1} < \infty, \\ Y_n^x & \text{if } \tau_{n+1} = \infty. \end{cases}$$

We set $\tau := \lim_{n \to \infty} \tau_n$ and

$$N := \inf\big\{n \geq 0 : \tau_{n+1} = \infty\big\} = \inf\big\{n \geq 0 : c(Y_n^x) = 0\big\}.$$

We claim that $(X_t^{s,x})_{t \geq s}$ defined as

$$X_t^x := \begin{cases} Y_k^x & \text{if } t \in [\tau_k, \tau_{k+1}),\ 0 \leq k < N + 1, \\ \infty & \text{if } t \geq \tau \end{cases}$$

solves the evolution equation (2.23). It is easy to see that $t \mapsto X_t$ is right-continuous with left limits, that $\tau = \inf\{r \geq s : X_r = \infty\}$ and $X_t = \infty$ for all $t \geq \tau$, and that (2.23) holds for all $t \neq \tau$. To see that (2.23) holds at the time τ should it be finite, we need to show that $X_{\tau-} = \infty$ on the event that $\tau < \infty$. To see this, imagine on the contrary that $\tau < \infty$ while X_t does not converge to ∞ as $t \to \tau$. By the definition of the one-point compactification, this implies that there exists a finite set $S' \subset S$ and times $s_n \to \tau$ such that $X_{s_n} \in S'$. This, in turn, implies that during the finite time interval $[0, \tau)$, the function X_t makes infinitely many jumps that start at some point in S' and end in some other point in S. But this is impossible, since by (2.20),

$$\bigcup_{y \in S'} \big\{t \in \mathbb{R} : \mathfrak{m}_t^\omega(y) \neq y\big\}$$

is a locally finite subset of $\mathbb{R}$. This completes the proof that $(X_t^{s,x})_{t\geq s}$ solves (2.23). If $(X'_t)_{t\geq s}$ is another solution, then we see by induction that $X'_t = X_t^{s,x}$ for all $0 \leq t \leq \tau_n$ and for all $n \geq 0$. This implies that $X'_{\tau-} = \infty$. By (2.23) and the way we have defined $m(\infty)$ for $m \in \mathcal{G}$, we must have $X'_\tau = X'_{\tau-}$ if $\tau < \infty$ so $X'_t = \infty$ for all $t \geq \tau$ by property (ii), which shows that solutions to (2.23) are unique.

Note that our previous argument holds almost surely for all $x \in S$ and $s \in \mathbb{R}$ simultaneously, that is, this includes random times s that may be chosen in dependence on the Poisson set ω. For deterministic x and s, we claim that $(X_{s+t}^{s,x})_{t\geq 0}$ is distributed as the continuous-time Markov chain with generator G and initial state x. To see this, let $(\sigma'_k)_{k\geq 0}$ be i.i.d. exponentially distributed random variables with mean one, independent of everything else. Define $(\sigma_k)_{k\geq 0}$ by $\sigma_k := c(Y_k^x)\tau_k$ if $c(Y_k^x) > 0$ and $\sigma_k := \sigma'_k$ otherwise. Using Proposition 2.19 we see by induction that $(Y_k^x)_{k\geq 0}$ is the embedded Markov chain, $(\sigma_k)_{k\geq 0}$ are i.i.d. exponentially distributed random variables with mean one, independent of $(Y_k^x)_{k\geq 0}$, and $(X_{s+t}^{s,x})_{t\geq 0}$ is the continuous-time Markov chain constructed in terms of its embedded Markov chain and exponential holding times as in Section 2.2.

Let $(\mathbf{X}_{s,t})_{s\leq t}$ be defined in (2.24). Then it is straightforward to check that $(\mathbf{X}_{s,t})_{s\leq t}$ is a stochastic flow in the sense of (2.25). If $X_0 = x$ is deterministic, then we have just seen that the process in (2.26) is distributed as the continuous-time Markov chain with generator G and initial law μ. The general case follows by conditioning on X_0, which is independent of everything else.

To complete the proof, we must show that if G is nonexplosive, then almost surely, $\mathbf{X}_{s,t}$ maps S into itself for all $s \leq t$. In other words, we must show that

$$\mathbf{X}_{s,t}(x) \in S \quad \forall s,t \in \mathbb{R} \text{ with } s \leq t \text{ and } x \in S \quad \text{a.s.}$$

If s is deterministic, then by what we have proved $X_t := \mathbf{X}_{s,s+t}(x)$ $(t \geq 0)$ is the continuous-time Markov chain with generator G and initial state x, so if this process is nonexplosive, then

$$\mathbf{X}_{s,t}(x) \in S \quad \forall t \in [s,\infty) \quad \text{a.s.} \quad (s \in \mathbb{R},\ x \in S).$$

Since $s \mapsto \mathbf{X}_{s,t}(x)$ is constant between the times of the Poisson point process

$$\{s \in \mathbb{R} : \mathfrak{m}_s^\omega(x) \neq x\}$$

which has finite intensity by (2.20), we can improve our previous statement to

$$\mathbf{X}_{s,t}(x) \in S \quad \forall s,t \in \mathbb{R} \text{ with } s \leq t \quad \text{a.s.} \quad (x \in S),$$

and since S is countable, we see that the statement holds for all $s,t \in \mathbb{R}$ with $s \leq t$ and $x \in S$ simultaneously. ∎

Exercise 2.23 *Use formula (2.26) as well as the fact that the stochastic flow $(\mathbf{X}_{s,t})_{s\leq t}$ is stationary with independent increments to give an alternative proof of the fact that the construction in Section 2.2 of $(X_t^x)_{t\geq 0}$ via the embedded Markov chain yields a Markov process in the sense of (2.5) with state space S_∞ and transition kernels $(\overline{P}_t)_{t\geq 0}$.*

For later use, we conclude this section with a theorem that is very similar to Theorem 2.22. The starting point is again a graphical representation ω that is associated with a random mapping representation of the generator G of a continuous-time Markov chain. The only difference is that this time, we will apply the maps associated with elements of ω in the reverse order. By definition, we say that a random function $X\colon (\infty, u] \to S_\infty$ solves the evolution equation

$$X_{t-} = \mathfrak{m}_t^{\omega}(X_t) \qquad (t \le u) \tag{2.27}$$

if

(i) $\lim_{r\downarrow t} X_r = X_t$ $(t < u)$ and $\lim_{r\uparrow t} X_r =: X_{t-}$ exists $(t \le u)$,
(ii) $X_{t-} = \infty$ for all $t \le \tau := \sup\{t \le u : X_t = \infty\}$,

and (2.27) holds. Note that we allow for the case that $(m, u) \in \omega$ for some $m \in \mathcal{G}$ and in this case, it is possible that $X_{u-} \ne X_u$. This is a difference with solutions of (2.23) which never make a jump at the initial time s. Because we work "backwards in time", in (2.30) we obtain a Markov process with left-continuous sample paths. In spite of these small differences, the proof of Theorem 2.24 is practically identical to the proof of Theorem 2.22.

Theorem 2.24 (Backward stochastic flow) *Let G be the generator of a continuous-time Markov chain with countable state space S and let ω be the graphical representation associated with a random mapping representation of G. Then almost surely, for each $u \in \mathbb{R}$ and $x \in S_\infty$, there exists a unique solution $(X_t^{u,x})_{t\le u}$ to the evolution equation (2.27) with final state $X_u^{u,x} = x$. Setting*

$$\mathbf{X}_{u,t}(x) := X_t^{u,x} \qquad (u \ge t,\ x \in S_\infty) \tag{2.28}$$

defines a collection of maps $(\mathbf{X}_{u,t})_{u\ge t}$ from S_∞ into itself such that

$$\mathbf{X}_{u,u} = 1 \quad \text{and} \quad \mathbf{X}_{t,s} \circ \mathbf{X}_{u,t} = \mathbf{X}_{u,s} \qquad (u \ge t \ge s). \tag{2.29}$$

If G is nonexplosive, then almost surely, $\mathbf{X}_{u,t}$ maps S into itself for all $u \ge t$. If $u \in \mathbb{R}$ and X_0 is an S-valued random variable with law μ, independent of ω, then the process $(X_t)_{t\ge 0}$ defined as

$$X_t := \mathbf{X}_{u,u-t}(X_0) \qquad (t \ge 0) \tag{2.30}$$

is distributed as the left-continuous modification of the continuous-time Markov chain with generator G and initial law μ.

We call a collection of maps $(\mathbf{X}_{u,t})_{u\ge t}$ as in (2.29) a *backward stochastic flow*. Stationarity and independent increments are defined as in the forward case.

Exercise 2.25 *Prove Theorem 2.24.*

2.7 An Example: ASEP

In this section, we look at an asymmetric simple exclusion process (ASEP) on the natural numbers with a finite, fixed number of particles. This demonstrates the theory developed so far and at the same time serves as a warm-up for the final two sections of this chapter, in which we show more generally how to construct interacting particle systems in which the lattice may be infinite, but the number of particles is finite. The construction of interacting particle systems with infinitely many particles will have to wait till Chapter 4.

We fix an integer $n \geq 1$ and write

$$\mathcal{S}_n := \big\{x \in \{0,1\}^{\mathbb{N}} : \sum_{i=0}^{\infty} x(i) = n\big\}.$$

For each $i, j \in \mathbb{N}$ with $i \neq j$, we let $\mathtt{asep}_{ij}$ denote the asymmetric exclusion map defined in Section 1.7. We observe that $\mathtt{asep}_{ij}$ preserves the number of particles, that is, it maps the space $\mathcal{S}_n$ into itself. We will be interested in the continuous-time Markov chain with countable state space $\mathcal{S}_n$ and generator

$$\begin{aligned} Gf(x) :=& \sum_{i=1}^{\infty} r_i^- \big\{f\big(\mathtt{asep}_{i,i-1}(x)\big) - f\big(x\big)\big\} \\ &+ \sum_{i=1}^{\infty} r_i^+ \big\{f\big(\mathtt{asep}_{i-1,i}(x)\big) - f\big(x\big)\big\}, \end{aligned} \tag{2.31}$$

where $r_i^\pm$ are nonnegative rates. To see that this is indeed the generator of a continuous-time Markov chain with state space $\mathcal{S}_n$, we must check condition (2.20) of Lemma 2.21. Since $\mathtt{asep}_{ij}(x) \neq x$ if and only if $x(i) = 1$ and $x(j) = 0$, the constant $c(x)$ from (2.20) is given by

$$c(x) = \sum_{i=1}^{\infty} 1_{\{x(i-1)=0,\ x(i)=1\}} r_i^- + \sum_{i=1}^{\infty} 1_{\{x(i-1)=1,\ x(i)=0\}} r_i^+. \tag{2.32}$$

For $x \in \mathcal{S}_n$, the sums in (2.32) have at most $2n$ nonzero terms so clearly $c(x) < \infty$ for all $x \in \mathcal{S}_n$. It follows that G, defined in (2.31) is the generator of a (possibly explosive) continuous-time Markov chain with state space $\mathcal{S}_n$.

Lemma 2.26 (Nonexplosiveness) *Assume that there exists a constant $K < \infty$ such that $r_i^+ \leq Ki$ $(i \geq 1)$. Then the continuous-time Markov chain $(X_t)_{t\geq 0}$ with generator G and state space $\mathcal{S}_n$ is nonexplosive.*

Proof For $x \in \mathcal{S}_n$, let $R(x) := \sup\{i \in \mathbb{N} : x(i) = 1\}$ denote the position of the right-most particle. We will apply Theorem 2.12 to the Lyapunov function

$$L(x) := \big(R(x) + 1\big)^2 \qquad (x \in \mathcal{S}_n).$$

Since L can only increase due to the right-most particle making a jump to the right,

$$\begin{aligned} GL(x) \leq& r^+_{R(x)+1}\big[\big(R(x)+2\big)^2 - \big(R(x)+1\big)^2\big] \\ \leq& K\big(R(x)+1\big)\big(2R(x)+3\big) \leq 3K\big(R(x)+1\big)^2, \end{aligned}$$

so condition (ii) of Theorem 2.12 is satisfied with $\lambda = 3K$. In view of (2.32), condition (i) is also satisfied, so we conclude that $(X_t)_{t\geq 0}$ is nonexplosive. ∎

Exercise 2.27 *Fix $\frac{1}{2} < p \leq 1$ and $\alpha > 1$ and assume that $r_i^- = (1-p)i^\alpha$ and $r_i^+ = p(i-1)^\alpha$ $(i \geq 1)$. Assume that $n = 1$ (there is only one particle) and write $X_t = e_{\xi_t}$ where $e_i \in \{0,1\}^\Lambda$ is defined as $e_i(j) := 1$ if $i = j$ and $:= 0$ otherwise. Then $(\xi_t)_{t\geq 0}$ is a continuous-time Markov chain with state space $\mathbb{N}$ that jumps from i to $i-1$ with rate $(1-p)i^\alpha$ $(i \geq 1)$ and from i to*

$i+1$ with rate pi^α ($i \geq 0$). Let $(Y_k)_{k\geq 0}$ be the embedded Markov chain of $(\xi_t)_{t\geq 0}$. Show that

$$\sum_{k=0}^{\infty} Y_k^{-\alpha} < \infty \quad \text{a.s.}$$

Use this to conclude that the continuous-time Markov chain $(X_t)_{t\geq 0}$ is explosive.

Formula (2.31) is a random mapping representation of the generator G. Combining Lemma 2.26 with Theorem 2.22, we can use this random mapping representation to define a stochastic flow $(\mathbf{X}_{s,t})_{s\leq t}$ on $\mathcal{S}_n$, and then construct the continuous-time Markov chain $(X_t)_{t\geq 0}$ in terms of this stochastic flow $(\mathbf{X}_{s,t})_{s\leq t}$ as in (2.26).

Random mapping representations are in general not unique, and different random mapping representations lead to different stochastic flows for the same continuous-time Markov chain. To demonstrate this, we assume from now on that

$$r_i := r_i^- = r_i^+ \qquad (i \geq 1). \tag{2.33}$$

We recall the definition of the (symmetric) exclusion map $\texttt{excl}_{ij}$ from Section 1.7. We claim that under the symmetry assumption (2.33), we can rewrite our definition of the generator G from (2.31) as

$$Gf(x) = \sum_{i=1}^{\infty} r_i \big\{ f\big(\texttt{excl}_{i-1,i}(x)\big) - f(x) \big\}. \tag{2.34}$$

To check this, it suffices to check that for each $x, y \in \mathcal{S}_n$ with $x \neq y$, the rate $G(x,y)$ of jumps from x to y is the same regardless of whether we define G by (2.31) or by (2.34). This rate is zero unless we are in one of the following two cases:

(i) There exists an $i \geq 1$ such that $x(i-1) = 0$, $x(i) = 1$, $y(i-1) = 1$, $y(i) = 0$, and $x(j) = y(j)$ for all $j \notin \{i-1, i\}$,
(ii) There exists an $i \geq 1$ such that $x(i-1) = 1$, $x(i) = 0$, $y(i-1) = 0$, $y(i) = 1$, and $x(j) = y(j)$ for all $j \notin \{i-1, i\}$.

Defining G as in (2.31), we see that $G(x,y) = r_i^-$ in case I and $G(x,y) = r_i^+$ in case II. On the other hand, defining G as in (2.34), we have $G(x,y) = r_i$ in both cases. In particular, under the symmetry assumption (2.33), both definitions are equivalent.

Even though under the assumption (2.33), formulas (2.31) and (2.34) define the same generator, they are different random mapping representations that lead to different stochastic flows. This is illustrated[7] in Figure 2.1. To further illuminate this, let us define

$$\mathcal{S}_{\text{fin}} := \bigcup_{n=0}^{\infty} \mathcal{S}_n.$$

Note that $\mathcal{S}_{\text{fin}}$ is countable. Our previous arguments show that G is the generator of a continuous-time Markov chain with state space $\mathcal{S}_{\text{fin}}$. By Theorem 2.22, we can use the random mapping representations (2.31) and (2.34) to construct two different stochastic flows

[7] In general, two different graphical representations of the same Markov process only yield two Markov processes that are equal in law. In Figure 2.1, for didactic purposes, these processes together with their graphical representations have been coupled so that the Markov processes are a.s. equal. In this example, can you think of a coupling that achieves this?

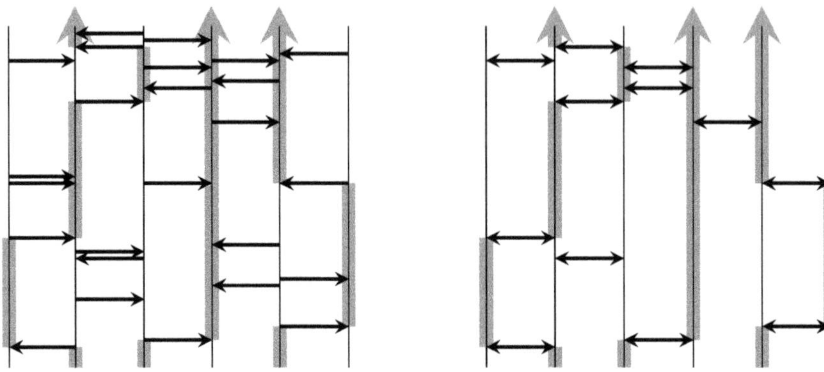

Figure 2.1 Two different Poisson constructions of the same continuous-time Markov chain. Each picture shows a different graphical representation of the same symmetric exclusion process. Time is plotted upwards. The arrows in the picture on the left symbolize the application of the map $\texttt{asep}_{ij}$, which has the effect that if there is a particle at i and the site j is vacant, then the particle at i jumps to j. The arrows in the picture on the right symbolize the application of the map $\texttt{excl}_{ij}$, which has the effect that particles at i and j exchange their positions. The Poisson density of arrows in the left picture is twice as high as in the right picture (or the same, if we count double arrows double).

$(\mathbf{X}_{s,t})_{s\le t}$ and $(\mathbf{X}'_{s,t})_{s\le t}$. The next exercise demonstrates that these stochastic flows have different properties.

Exercise 2.28 *Show that the random maps* $\mathbf{X}'_{s,t}: \mathcal{S}_{\rm fin} \to \mathcal{S}_{\rm fin}$ *are* additive *in the sense that*

$$\mathbf{X}'_{s,t}(x \vee y) = \mathbf{X}'_{s,t}(x) \vee \mathbf{X}'_{s,t}(y) \qquad (s \le t,\ x, y \in \mathcal{S}_{\rm fin}),$$

where $(x \vee y)(i) = x(i) \vee y(i)$ *denotes the pointwise maximum of two configurations* $x, y \in \mathcal{S}_{\rm fin}$. *Show that the maps* $\mathbf{X}_{s,t}: \mathcal{S}_{\rm fin} \to \mathcal{S}_{\rm fin}$ *are not additive.*

Exercise 2.29 *Show that if (2.33) is weakened to* $r_i^- \le r_i^+$ $(i \ge 1)$, *then the generator* G *in (2.31) can be rewritten as*

$$\begin{aligned} Gf(x) &= \sum_{i=1}^{\infty} r_i^- \{ f(\texttt{excl}_{i-1,i}(x)) - f(x) \} \\ &\quad + \sum_{i=1}^{\infty} (r_i^+ - r_i^-) \{ f(\texttt{asep}_{i-1,i}(x)) - f(x) \}. \end{aligned}$$

2.8 Local Maps

In this section, we return to the general setup of Section 1.1. Thus, S is a finite set, called the *local state space*, Λ is a countable set, called the *lattice*, and S^Λ denotes the Cartesian product of Λ copies of S, that is, this is the space of all functions $x: \Lambda \to S$. The interacting particle systems we are interested in are Markov processes with state space S^Λ and generator

G of the form (1.1), that is,

$$Gf(x) = \sum_{m \in \mathcal{G}} r_m \{ f(m(x)) - f(x) \},$$

were $\mathcal{G}$ is a set whose elements are maps $m \colon S^\Lambda \to S^\Lambda$ and $(r_m)_{m \in \mathcal{G}}$ are nonnegative rates.

If the lattice Λ is finite, then so is S^Λ, and we can immediately apply Theorem 2.22 to construct our interacting particle system from a graphical representation ω. If Λ is infinite, then S^Λ is uncountable (as long as S has at least two elements), so Theorem 2.22 is not applicable. In Chapter 4, we will develop the methods needed to construct interacting particle systems on infinite lattices from a graphical representation. There is one important special case where Theorem 2.22 is applicable, however, even if the lattice is infinite. Many interacting particle systems with a local state space of the form $S = \{0, 1\}$ have the property that if we start the system in an initial configuration with finitely many ones, then the system stays in such states for all times. In the previous section, we have shown this for the exclusion process. Other examples are the voter model, the contact process, and various systems of branching and coalescing particles. A counterexample is formed by stochastic Ising models.

In Section 2.9, we will show how Theorem 2.22 can be applied to construct interacting particle systems on infinite lattices, but started in an initial configuration with finitely many ones. To prepare for this, in the present section, we take a closer look at the sort of maps that are typically used to construct an interacting particle system, such as the voter map in (1.4) and the branching and death maps in (1.6) and (1.7).

We will always equip the state space S^Λ of an interacting particle system with the *product topology*, which says that a sequence $x_n \in S^\Lambda$ converges to a limit x if and only if

$$x_n(i) \underset{n \to \infty}{\longrightarrow} x(i) \qquad \forall i \in \Lambda.$$

Note that since S is finite, this simply says that for each $i \in \Lambda$, there is an N (which may depend on i) such that $x_n(i) = x(i)$ for all $n \geq N$. Since S is finite, it is in particular compact, so by Tychonoff's theorem, the space S^Λ is compact in the product topology.

Let S and T be finite sets, let Λ be a countable set, and let $f \colon S^\Lambda \to T$ be a function. Then we say that a point $j \in \Lambda$ is *f-relevant* if

$$\exists x, y \in S^\Lambda \text{ s.t. } f(x) \neq f(y) \text{ and } x(k) = y(k)\ \forall k \neq j,$$

that is, changing the value of x in j may change the value of $f(x)$. We write

$$\mathcal{R}(f) := \{ j \in \Lambda : j \text{ is } f\text{-relevant} \}.$$

The following lemma (which we have taken from [SS18, Lemma 24]) says that a function $f \colon S^\Lambda \to T$ is continuous with respect to the product topology if and only if it depends on finitely many coordinates.

Lemma 2.30 (Continuous maps) *Let S and T be finite sets and let Λ be a countable set. Then a function $f \colon S^\Lambda \to T$ is continuous with respect to the product topology if and only if the following two conditions are satisfied:*

(i) *$\mathcal{R}(f)$ is finite,*
(ii) *If $x, y \in S^\Lambda$ satisfy $x(j) = y(j)$ for all $j \in \mathcal{R}(f)$, then $f(x) = f(y)$.*

Before we give the proof of Lemma 2.30, we first make some observations. The following exercise shows how continuity can fail if condition (i) of Lemma 2.30 does not hold.

Exercise 2.31 (A discontinuous map) *Let* $2\mathbb{N} := \{2n : n \in \mathbb{N}\}$ *and* $2\mathbb{N}+1 := \{2n+1 : n \in \mathbb{N}\}$. *Define* $f\colon \{0,1\}^{\mathbb{N}} \to \{0,1\}$ *by*

$$f(x) := \begin{cases} 1 & \text{if } \inf\{i \in \mathbb{N} : x(i) = 1\} \in 2\mathbb{N} \cup \{\infty\}, \\ 0 & \text{if } \inf\{i \in \mathbb{N} : x(i) = 1\} \in 2\mathbb{N}+1. \end{cases} \tag{2.35}$$

Show that f satisfies condition (ii) of Lemma 2.30 but not condition (i). Show that f is not continuous.

The following exercise shows that, contrary to what one might initially have guessed, condition (ii) of Lemma 2.30 is not automatically satisfied, even when condition (i) holds.

Exercise 2.32 (Another discontinuous map) *Define* $f\colon \{0,1\}^{\mathbb{N}} \to \{0,1\}$ *by*

$$f(x) := \begin{cases} 1 & \text{if } \{i \in \mathbb{N} : x(i) = 1\} \text{ is finite}, \\ 0 & \text{if } \{i \in \mathbb{N} : x(i) = 1\} \text{ is infinite}. \end{cases} \tag{2.36}$$

Show that f satisfies condition (i) of Lemma 2.30 but not condition (ii). Show that f is not continuous.

Proof of Lemma 2.30 Let $(\alpha_j)_{j\in\Lambda}$ be strictly positive constants such that $\sum_{j\in\Lambda}\alpha_j < \infty$. Then the metric

$$d(x,y) := \sum_{j\in\Lambda} \alpha_j 1_{\{x(j) \neq y(j)\}} \qquad (x, y \in S^\Lambda) \tag{2.37}$$

generates the product topology on S^Λ. By Tychonoff's theorem, S^Λ is compact, so the function f is uniformly continuous. Since the target space T is finite, this means that there exists an $\varepsilon > 0$ such that $d(x,y) < \varepsilon$ implies $f(x) = f(y)$. Since $\sum_{j\in\Lambda}\alpha_j < \infty$, there exists some finite $\Lambda' \subset \Lambda$ such that $\sum_{j\in\Lambda\setminus\Lambda'}\alpha_j < \varepsilon$. It follows that

(ii)' If $x, y \in S^\Lambda$ satisfy $x(j) = y(j)$ for all $j \in \Lambda'$, then $f(x) = f(y)$.

We conclude from this that $\mathcal{R}(f) \subset \Lambda'$, proving (i). If this is a strict inclusion, then we can inductively remove those points from Λ' that are not elements of $\mathcal{R}(f)$ while preserving the property (ii)', until in a finite number of steps we see that (ii) holds.

Conversely, if (i) and (ii) hold and $x_k \to x$ pointwise, then by (i) there exists some n such that $x_k(j) = x(j)$ for all $j \in \mathcal{R}(f)$ and hence by (ii) $f(x_k) = f(x)$ for all $k \geq n$, proving that f is continuous. ∎

For any map $m\colon S^\Lambda \to S^\Lambda$ and $i \in \Lambda$, we define $m[i]\colon S^\Lambda \to S$ by $m[i](x) := m(x)(i)$ $(x \in S^\Lambda,\ i \in \Lambda)$. It follows immediately from the definition of the product topology that m is continuous if and only if $m[i]$ is continuous for all $i \in \Lambda$. We let

$$\mathcal{D}(m) := \big\{ i \in \Lambda : \exists x \in S^\Lambda \text{ s.t. } m(x)(i) \neq x(i) \big\}$$

denote the set of lattice points i for which $m[i]$ is not the trivial map $m[i](x) = x(i)$ $(x \in S^\Lambda)$. Note that $\mathcal{R}(m[i]) = \{i\}$ if $i \notin \mathcal{D}(m)$.

By definition, a *local map* is a function $m\colon S^\Lambda \to S^\Lambda$ such that:

(i) m is continuous,
(ii) $\mathcal{D}(m)$ is finite.

In view of Lemma 2.30, this says that m is local if m changes the values of at most finitely many lattice points using information from finitely many lattice points only. The following exercise describes yet another way to look at local maps.

Exercise 2.33 (Local maps) *Show that a map $m\colon S^\Lambda \to S^\Lambda$ is local if and only if there exists a finite set $\Delta \subset \Lambda$ and a map $m'\colon S^\Delta \to S^\Delta$ such that*

$$m(x)(k) = \begin{cases} m'\big((x(i))_{i\in\Delta}\big)(k) & \text{if } k \in \Delta, \\ x(k) & \text{otherwise.} \end{cases}$$

Before we continue, it is good to see a number of examples.

- The voter map $\texttt{vot}_{ij}$ defined in (1.4) satisfies

$$\mathcal{D}(\texttt{vot}_{ij}) = \{j\} \quad \text{and} \quad \mathcal{R}(\texttt{vot}_{ij}[j]) = \{i\},$$

  since only the type at j changes, and it suffices to know the type at i to predict the new type at j.
- The branching map $\texttt{bra}_{ij}$ defined in (1.6) satisfies

$$\mathcal{D}(\texttt{bra}_{ij}) = \{j\} \quad \text{and} \quad \mathcal{R}(\texttt{bra}_{ij}[j]) = \{i, j\},$$

  since only the type at j changes, but we need to know both the type at i and at j to predict the new type at j since $\texttt{bra}_{ij}(x)(j) = x(i) \vee x(j)$.
- The death map $\texttt{death}_i$ defined in (1.7) satisfies

$$\mathcal{D}(\texttt{death}_i) = \{i\} \quad \text{and} \quad \mathcal{R}(\texttt{death}_i[i]) = \emptyset$$

  since only the type at i changes, and the new type at i is 0 regardless of x.
- For each $i \in \Lambda$, we can similarly define a *birth map* $\texttt{birth}_i\colon \{0,1\}^\Lambda \to \{0,1\}^\Lambda$ as

$$\texttt{birth}_i(x)(k) := \begin{cases} 1 & \text{if } k = i, \\ x(k) & \text{otherwise.} \end{cases} \tag{2.38}$$

  Then

$$\mathcal{D}(\texttt{birth}_i) = \{i\} \quad \text{and} \quad \mathcal{R}(\texttt{birth}_i[i]) = \emptyset.$$

- The coalescing random walk map $\texttt{rw}_{ij}$ defined in (1.24) satisfies

$$\mathcal{D}(\texttt{rw}_{ij}) = \{i, j\}, \quad \mathcal{R}(\texttt{rw}_{ij}[i]) = \emptyset, \quad \text{and} \quad \mathcal{R}(\texttt{rw}_{ij}[j]) = \{i, j\},$$

  since the types at both i and j can change, the new type at i is 0 regardless of the previous state, but to calculate $\texttt{rw}_{ij}(x)(j)$, we need to know both $x(i)$ and $x(j)$.

Exercise 2.34 (Exclusion and cooperative branching maps) *Recall the asymmetric and symmetric exclusion maps $\texttt{asep}_{ij}$ and $\texttt{excl}_{ij}$ defined in (1.20) and (1.22), and the cooperative branching map $\texttt{coop}_{ii'j}$ defined in (1.27). When m is any of these maps, determine $\mathcal{D}(m)$, and determine $\mathcal{R}(m[i])$ for all $i \in \mathcal{D}(m)$.*

2.9 Systems of Finitely Many Particles

Throughout this section, we assume that S is a finite set containing a special element that we denote by 0. For $x \in S^\Lambda$, we introduce the notation

$$|x| := \big|\{i \in \Lambda : x(i) \neq 0\}\big| \qquad (x \in S^\Lambda)$$

and we write

$$\mathcal{S}(\Lambda) := S^\Lambda \quad \text{and} \quad \mathcal{S}_{\text{fin}}(\Lambda) := \big\{x \in \mathcal{S}(\Lambda) : |x| < \infty\big\}.$$

It is easy to see that $\mathcal{S}_{\text{fin}}(\Lambda)$ is countable. We let $\underline{0} \in \mathcal{S}(\Lambda)$ denote the configuration that is identically zero, that is, this is the constant function defined as $\underline{0}(i) := 0$ $(i \in \Lambda)$. We will be interested in local maps $m\colon \mathcal{S}(\Lambda) \to \mathcal{S}(\Lambda)$ that satisfy

$$m(\underline{0}) = \underline{0}. \tag{2.39}$$

Almost all the examples of local maps mentioned in the previous section satisfy (2.39). Indeed, this holds for the local maps

$$\texttt{vot}_{ij}, \quad \texttt{bra}_{ij}, \quad \texttt{death}_i, \quad \texttt{rw}_{ij}, \quad \texttt{asep}_{ij}, \quad \texttt{excl}_{ij}, \quad \text{and} \quad \texttt{coop}_{ijk},$$

while $\texttt{birth}_i$ is the only local map mentioned in the previous section that does not map $\underline{0}$ into itself.

Let $\mathcal{G}$ be a countable collection of local maps $m\colon \mathcal{S}(\Lambda) \to \mathcal{S}(\Lambda)$ that all satisfy (2.39) and let $(r_m)_{m\in\mathcal{G}}$ be rates. Then, under suitable assumptions on the rates, we may expect that

$$Gf(x) := \sum_{m\in\mathcal{G}} r_m\big\{f\big(m(x)\big) - f\big(x\big)\big\} \tag{2.40}$$

is the generator of a nonexplosive continuous-time Markov chain with countable state space $\mathcal{S}_{\text{fin}}(\Lambda)$. To formulate sufficient conditions for this to be true, for any local map $m\colon S^\Lambda \to S^\Lambda$ we introduce the following notation:

$$\mathcal{R}(m) := \big\{(i,j) \in \Lambda^2 : i \text{ is } m[j]\text{-relevant}\big\},$$
$$\mathcal{R}^\uparrow_i(m) := \big\{j \in \Lambda : (i,j) \in \mathcal{R}(m)\big\}, \quad \mathcal{R}^\downarrow_j(m) := \big\{i \in \Lambda : (i,j) \in \mathcal{R}(m)\big\}.$$

Here is the main result of this section. The form of condition (2.41) is inspired by [Lat24]. Below, we let 1_A denote the indictor function of a set $A \subset \Lambda$, that is, $1_A(i) := 1$ if $i \in A$ and $:= 0$ if $i \in \Lambda\backslash A$.

Theorem 2.35 (Finite particle configurations) *Let S and Λ be a finite and countable set, respectively, and assume that S contains a special element denoted as 0. Let $\mathcal{G}$ be a countable collection of local maps $m\colon S^\Lambda \to S^\Lambda$ such that $m(\underline{0}) = \underline{0}$ for all $m \in \mathcal{G}$ and let $(r_m)_{m\in\mathcal{G}}$ be nonnegative rates. Assume that*

$$\text{(i)}\ \sup_{i\in\Lambda}\sum_{m\in\mathcal{G}} r_m 1_{\mathcal{D}(m)}(i) < \infty, \quad \text{(ii)}\ \sup_{i\in\Lambda}\sum_{m\in\mathcal{G}} r_m\big|\mathcal{R}^\uparrow_i(m)\backslash\{i\}\big| < \infty. \tag{2.41}$$

Then G defined in (2.40) is the generator of a nonexplosive continuous-time Markov chain with state space $\mathcal{S}_{\text{fin}}(\Lambda)$. Moreover, the process started in $X_0 = x \in \mathcal{S}_{\text{fin}}(\Lambda)$ satisfies

$$\mathbb{E}^x\big[|X_t|\big] \le |x| e^{K_\uparrow t} \quad (t \ge 0) \quad \textit{with} \quad K_\uparrow := \sup_{i\in\Lambda}\sum_{m\in\mathcal{G}} r_m\big(|\mathcal{R}^\uparrow_i(m)| - 1\big). \tag{2.42}$$

Proof We start by checking condition (2.20), which is necessary and sufficient for (2.40) to define the generator of a (possibly explosive) continuous-time Markov chain. In our present setting, (2.20) reads

$$\sum_{m:\, m(x)\neq x} r_m < \infty \quad \text{for all } x \in \mathcal{S}_{\text{fin}}(\Lambda).$$

Let $x \in \mathcal{S}_{\text{fin}}(\Lambda)$ and let $A := \{i \in \Lambda : x(i) \neq 0\}$, which is finite by the definition of $\mathcal{S}_{\text{fin}}(\Lambda)$. If $m(x) \neq x$, then $m(x)(j) \neq x(j)$ for some some $j \in \mathcal{D}(m)$. If $j \notin A$, then by the fact that $m(\underline{0}) = \underline{0}$, there must exist an $i \in A$ such that $i \in \mathcal{R}(m[j])$. This allows us to estimate

$$\begin{aligned}\sum_{m:\, m(x)\neq x} r_m &\leq \sum_{j\in A}\sum_{m\in\mathcal{G}} 1_{\mathcal{D}(m)}(j) r_m + \sum_{j\in\Lambda\backslash A}\sum_{i\in A}\sum_{m\in\mathcal{G}} 1_{\mathcal{R}^{\uparrow}_i(m)}(j) r_m \\ &= \sum_{i\in A}\Big(\sum_{m\in\mathcal{G}} 1_{\mathcal{D}(m)}(i) r_m + \sum_{m\in\mathcal{G}} |\mathcal{R}^{\uparrow}_i(m)\backslash A| r_m\Big),\end{aligned}$$

which is finite by (2.41).

It remains to prove that G is nonexplosive. We apply Theorem 2.12 to the Lyapunov function

$$L(x) := |x| \quad \big(x \in \mathcal{S}_{\text{fin}}(\Lambda)\big). \tag{2.43}$$

Note that $c(x) := -G(x,x) = \sum_{m:\, m(x)\neq x} r_m$, so our previous calculation shows that the function L satisfies condition (i) of Theorem 2.12. It remains to check condition (ii). Since each $m \in \mathcal{G}$ satisfies $m(\underline{0}) = \underline{0}$, if $m(x)(j) \neq 0$ for some $j \in \Lambda$, then there must be an $i \in \Lambda$ such that $x(i) \neq 0$ and $j \in \mathcal{R}^{\uparrow}_i(m)$, which allows us to estimate

$$|m(x)| = \sum_{i:\, x(i)\neq 0}\sum_{j} 1_{\mathcal{R}^{\uparrow}_i(m)}(j) = \sum_{i:\, x(i)\neq 0} \big|\mathcal{R}^{\uparrow}_i(m)\big|.$$

It follows that

$$\begin{aligned}GL(x) &= \sum_{m\in\mathcal{G}} r_m\big\{L\big(m(x)\big) - L\big(x\big)\big\} \\ &\leq \sum_{m\in\mathcal{G}} r_m \sum_{i:\, x(i)\neq 0} \big(|\mathcal{R}^{\uparrow}_i(m)| - 1\big) = \sum_{i:\, x(i)\neq 0}\sum_{m\in\mathcal{G}} r_m\big(|\mathcal{R}^{\uparrow}_i(m)| - 1\big) \\ &\leq L(x) \sup_{i\in\Lambda} \sum_{m\in\mathcal{G}} r_m\big(|\mathcal{R}^{\uparrow}_i(m)| - 1\big) \qquad (x \in \mathcal{S}_{\text{fin}}(\Lambda)).\end{aligned}$$

This shows that condition (ii) of Theorem 2.12 is satisfied with $\lambda = K_{\uparrow}$, so G is nonexplosive. Formula (2.42) now follows from the exponential bound in Theorem 2.12. ∎

It is instructive to see some concrete examples of interacting particle systems to which Theorem 2.35 is applicable. Generalizing (1.5), if $\lambda \colon \Lambda^2 \to [0,\infty)$ is a function, then we can define a *voter model* generator by

$$G_{\text{vot}} f(x) := \sum_{i,j\in\Lambda^2} \lambda(i,j)\big\{f\big(\texttt{vot}_{ij}(x)\big) - f\big(x\big)\big\} \qquad (x \in S^{\Lambda}), \tag{2.44}$$

where $\lambda(i,j) \geq 0$ is the Poisson rate at which site j adopts the type of site i.

Exercise 2.36 *Show that the generator G_{vot} satisfies the assumptions of Theorem 2.35 if*

$$\sup_{i\in\Lambda}\Big[\sum_{j\in\Lambda}\lambda(j,i) + \sum_{j\in\Lambda}\lambda(i,j)\Big] < \infty.$$

Similarly, generalizing (1.8), we can define a *contact process* generator by

$$\begin{aligned} G_{\mathrm{cont}}f(x) := \sum_{i,j\in\Lambda} \lambda(i,j)\big\{f\big(\mathtt{bra}_{ij}(x)\big) - f\big(x\big)\big\} \\ + \delta \sum_{i\in\Lambda} \big\{f\big(\mathtt{death}_i(x)\big) - f\big(x\big)\big\}, \end{aligned} \tag{2.45}$$

where $\lambda(i, j) \geq 0$ is the *infection rate* from i to j and $\delta \geq 0$ is the *death rate.*

Exercise 2.37 *Show that the generator G_{cont} satisfies the assumptions of Theorem 2.35 if*

$$\sup_{i\in\Lambda} \Big[\sum_{j\in\Lambda} \lambda(j,i) + \sum_{j\in\Lambda} \lambda(i,j)\Big] < \infty.$$

For processes with a sort of translation invariant structure,[8] the expression $\sum_{j\in\Lambda} \lambda(j,i) + \sum_{j\in\Lambda} \lambda(i,j)$ does not depend on $i \in \Lambda$. Using this, one can check that for translation-invariant voter models, the condition in Exercise 2.36 is optimal. Indeed, if we start the process with a single one at i, then $\sum_{j\in\Lambda} \lambda(j,i)$ is the rate at which this one becomes a zero while $\sum_{j\in\Lambda} \lambda(i,j)$ is the rate at which this one produces another one somewhere. In the case of the contact process, we can actually do a bit better than Exercise 2.37.

Proposition 2.38 (Finite contact processes) *Assume that*

$$r := \sup_{i\in\Lambda} \sum_{j\in\Lambda} \lambda(i,j) < \infty. \tag{2.46}$$

Then G_{cont}, defined in (2.45), is the generator of a nonexplosive continuous-time Markov chain with state space $\mathcal{S}_{\mathrm{fin}}(\Lambda)$. Moreover,

$$\mathbb{E}^x\big[|X_t|\big] \leq e^{(r-\delta)t}|x| \qquad \big(t \geq 0,\ x \in \mathcal{S}_{\mathrm{fin}}(\Lambda)\big). \tag{2.47}$$

Proof For any $x \in \mathcal{S}_{\mathrm{fin}}(\Lambda)$, we can estimate the quantity $c(x)$ from (2.20) by

$$\sum_{m:\, m(x)\neq x} r_m = \delta|x| + \sum_{i:\, x(i)=1}\ \sum_{j:\, x(j)=0} \lambda(i,j) \leq (\delta + r)|x|,$$

where r is the quantity in (2.46). Since this is finite for each $x \in \mathcal{S}_{\mathrm{fin}}(\Lambda)$, G_{cont} is the generator of a (possibly explosive) continuous-time Markov chain with state space $\mathcal{S}_{\mathrm{fin}}(\Lambda)$.

To see that G_{cont} is nonexplosive, we apply Theorem 2.12 to the Lyapunov function in (2.43). Our previous calculation shows that L satisfies condition (i) of Theorem 2.12 so it remains to check condition (ii). We estimate

$$\begin{aligned} GL(x) &= \sum_{m\in\mathcal{G}} r_m\big\{L\big(m(x)\big) - L\big(x\big)\big\} \\ &= \sum_{i,j\in\Lambda} \lambda(i,j) 1_{\{x(i)=1,\ x(j)=0\}} - \delta \sum_{i\in\Lambda} 1_{\{x(i)=1\}} \leq (r-\delta)|x|, \end{aligned}$$

[8] To formalize this, let us call a bijection $\psi : \Lambda \to \Lambda$ such that $\lambda\big(\psi(i), \psi(j)\big) = \lambda(i, j)$ for all $i, j \in \Lambda$ an *automorphism* of λ. In analogy with the terminology for graphs, we can define λ to be *vertex transitive* if for each $i, j \in \Lambda$, there exists an automorphism ψ such that $\psi(i) = j$. If λ is vertex transitive, then $\sum_{j\in\Lambda} \lambda(j,i) + \sum_{j\in\Lambda} \lambda(i,j)$ does not depend on $i \in \Lambda$.

from which we see that condition (ii) of Theorem 2.12 is satisfied with $\lambda = r - \delta$. Theorem 2.12 now tells us that G_{cont} is nonexplosive and (2.47) holds. ∎

In particular, Proposition 2.38 tells us that if $r < \delta$, then the contact process *dies out* in the sense that

$$\mathbb{P}^x\big[X_t = \underline{0}\big] \underset{t\to\infty}{\longrightarrow} 1 \qquad \forall x \in \mathcal{S}_{\text{fin}}(\Lambda).$$

This is quite natural since r, defined in (2.46), is the maximal reproduction rate of an individual (assuming all other sites are vacant). If this is less than the death rate, then each individual produces on average less than one offspring before it dies, leading to an exponential decay of the population size.

3

The Mean-Field Limit

3.1 Processes on the Complete Graph

In Chapter 1, we have made acquaintances with a number of interacting particle systems. While some properties of these systems turn out to be easy to prove, other seemingly elementary questions can sometimes be remarkably difficult. A few examples of such hard problems have been mentioned in Chapter 1. In view of this, interacting particle systems are being studied by a range of different methods, from straightforward numerical simulations, as we have seen in Chapter 1, to nonrigorous renormalization group techniques and rigorous mathematical methods. All these approaches complement each other. In addition, when a given problem appears too hard, one often looks for simpler models that (one hopes) still catch the essence, or at least some essential features, of the behavior that one is interested in.

A standard way to turn a difficult model into an (often) much easier model is to take the *mean-field limit*, which we explain in the present chapter. Basically, this means that one replaces the graph structure of the underlying lattice that one is really interested in (in practice often $\mathbb{Z}^d$) by the structure of the complete graph with N vertices and then takes the limit $N \to \infty$. As we will see, many properties of "real" interacting particle systems are already reflected in these mean-field models. In particular, phase transitions can often already be observed, and even the values of critical exponents of high-dimensional models are correctly predicted by the mean-field model. In view of this, studying the mean-field limit is a wise first step in the study of any more complicated model that one may encounter.

Of course, not all phenomena can be captured by replacing the graph structure that one is really interested in by the complete graph. Comparing the real model with the mean-field model, one can learn which elements of the observed behavior are a consequence of the specific spatial structure of the lattice and which are not. Also for this reason, studying the mean-field limit should be part of a complete study of any interacting particle system.

3.2 The Mean-Field Limit of the Ising Model

In this section, we study the mean-field Ising model, also known as the *Curie–Weiss model*, with Glauber dynamics. We recall from formulas (1.11) and (1.12) in Chapter 1 that the Ising model is an interacting particle system with local state space $S = \{-1, +1\}$, where each site i updates its spin value $x(i) \in \{-1, +1\}$ at rate one. When a spin value is updated, the probability that the new value is $+1$, respectively, -1, is proportional to $e^{\beta N_{x,i}(+1)}$, respectively, $e^{\beta N_{x,i}(-1)}$, where $N_{x,i}(\sigma) := \sum_{j \in N_i} 1_{\{x(j)=\sigma\}}$ denotes the number of neighboring sites that have the spin value σ.

For the aim of taking the mean-field model, it will be convenient to formulate the model slightly differently. We let

$$\overline{N}_{x,i} := \frac{1}{|\mathcal{N}_i|} \sum_{j\in\mathcal{N}_i} 1_{\{x(j)=\sigma\}}$$

denote the fraction of neighbors that have the spin value σ, and consider the model where (compare (1.12))

$$\text{site } i \text{ flips to the value } \sigma \text{ with rate} \quad \frac{e^{\beta \overline{N}_{x,i}(\sigma)}}{\sum_{\tau\in\{-1,+1\}} e^{\beta \overline{N}_{x,i}(\tau)}}. \tag{3.1}$$

Assuming that $|\mathcal{N}_i|$ is just a constant that does not depend on $i \in \Lambda$ (as is the case for the model on $\mathbb{Z}^d$ and more generally on any vertex transitive graph), this is just a reparametrization of the original model where the parameter β is replaced by $\beta/|\mathcal{N}_i|$.

We now wish to construct the mean-field model, that is, the model on a complete graph Λ_N with $|\Lambda_N| = N$ vertices (sites), where each site is a neighbor of each other site. For mathematical simplicity, we even count a site as a neighbor of itself, that is, we set

$$\mathcal{N}_i := \Lambda_N \quad \text{and} \quad |\mathcal{N}_i| = N.$$

A consequence of this choice is that the *average magnetization*

$$\overline{X}_t := \frac{1}{N} \sum_{i\in\Lambda_N} X_t(i) \qquad (t \geq 0)$$

forms a Markov process $\overline{X} = (\overline{X}_t)_{t\geq 0}$. Indeed, $\overline{X}_t$ takes values in the space

$$\big\{ -1, -1 + \tfrac{2}{N}, \ldots, 1 - \tfrac{2}{N}, 1 \big\},$$

and jumps

$$\begin{aligned}
\overline{x} &\mapsto \overline{x} + \tfrac{2}{N} \qquad \text{with rate} \qquad N_x(-1) \frac{e^{\beta N_x(+1)/N}}{e^{\beta N_x(-1)/N} + e^{\beta N_x(+1)/N}},\\
\overline{x} &\mapsto \overline{x} - \tfrac{2}{N} \qquad \text{with rate} \qquad N_x(+1) \frac{e^{\beta N_x(-1)/N}}{e^{\beta N_x(-1)/N} + e^{\beta N_x(+1)/N}},
\end{aligned}$$

where $N_x(\sigma) := N_{x,i}(\sigma) = \sum_{j\in\Lambda_n} 1_{\{x(j)=\sigma\}}$ does not depend on $i \in \Lambda_N$. We observe that

$$N_x(+1)/N = (1 + \overline{x})/2 \quad \text{and} \quad N_x(-1)/N = (1 - \overline{x})/2.$$

In view of this, we can rewrite the jump rates of $\overline{X}$ as

$$\begin{aligned}
\overline{x} &\mapsto \overline{x} + \tfrac{2}{N} \qquad \text{with rate} \qquad N(1 - \overline{x})/2 \, \frac{e^{\beta(1+\overline{x})/2}}{e^{\beta(1-\overline{x})/2} + e^{\beta(1+\overline{x})/2}},\\
\overline{x} &\mapsto \overline{x} - \tfrac{2}{N} \qquad \text{with rate} \qquad N(1 + \overline{x})/2 \, \frac{e^{\beta(1-\overline{x})/2}}{e^{\beta(1-\overline{x})/2} + e^{\beta(1+\overline{x})/2}}.
\end{aligned}$$

In particular, since these rates are a function of $\overline{x}$ only (and do not depend on other functions of $x = (x(i))_{i\in\Lambda_N}$), we see that $\overline{X} = (\overline{X}_t)_{t\geq 0}$, on its own, is a Markov process. (This argument will be made rigorous in Section 3.4.) Canceling a common factor $e^{\beta/2}$ in the numerator and denominator of the rates, we can simplify our formulas a bit to

$$\begin{aligned} \overline{x} &\mapsto \overline{x} + \tfrac{2}{N} \quad \text{with rate} \quad r_+^N(\overline{x}) := N(1-\overline{x})/2\,\frac{e^{\beta\overline{x}/2}}{e^{-\beta\overline{x}/2} + e^{\beta\overline{x}/2}}, \\ \overline{x} &\mapsto \overline{x} - \tfrac{2}{N} \quad \text{with rate} \quad r_-^N(\overline{x}) := N(1+\overline{x})/2\,\frac{e^{-\beta\overline{x}/2}}{e^{-\beta\overline{x}/2} + e^{\beta\overline{x}/2}}. \end{aligned} \tag{3.2}$$

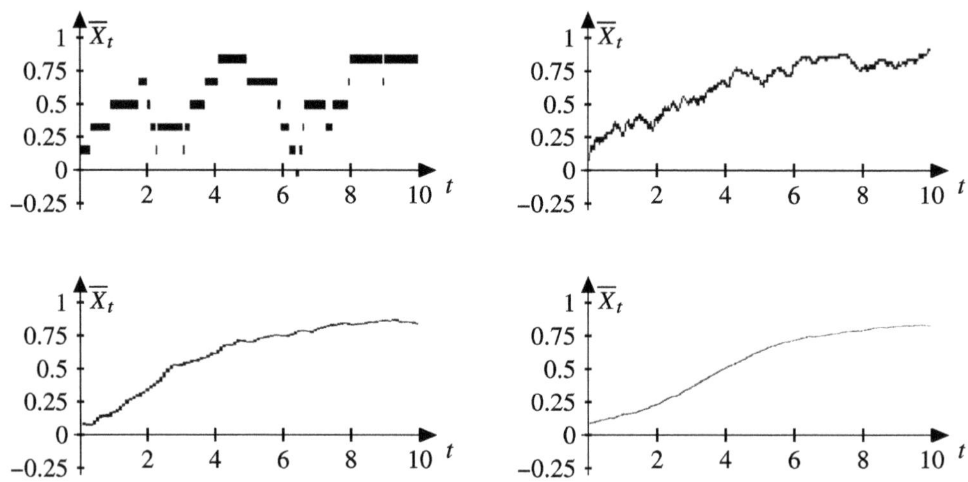

Figure 3.1 The mean-field Ising model on lattice with $N = 10$, 100, 1,000, and 10,000 sites, respectively. In these simulations, the parameter is $\beta = 3$, and the initial state is $\overline{X}_0 = 0.1$, except in the first picture, where $\overline{X}_0 = 0.2$.

In Figure 3.1, we can see simulations of the Markov process in (3.2) on a lattice with $N = 10$, 100, 1,000, and 10,000 sites, respectively. It appears that in the limit $N \to \infty$, the process $\overline{X}_t$ is given by a smooth, deterministic function.

It is not hard to guess what this function is. Indeed, denoting the generator of the process in (3.2) by $\overline{G}_{N,\beta}$, we see that

$$\mathbb{E}^{\overline{x}}[\overline{X}_t] = \overline{x} + tg_\beta(\overline{x}) + O(t^2) \quad \text{where} \quad g_\beta(\overline{x}) := \overline{G}_{N,\beta} f(\overline{x}) \quad \text{with} \quad f(\overline{x}) := \overline{x}.$$

We call the function g_β the *local drift* of the process $\overline{X}$. We calculate

$$\begin{aligned} g_\beta(\overline{x}) &= r_+^N(\overline{x}) \cdot \frac{2}{N} - r_-^N(\overline{x}) \cdot \frac{2}{N} = \frac{(1-\overline{x})e^{\beta\overline{x}/2} - (1+\overline{x})e^{-\beta\overline{x}/2}}{e^{\beta\overline{x}/2} + e^{-\beta\overline{x}/2}} \\ &= \frac{e^{\beta\overline{x}/2} - e^{-\beta\overline{x}/2}}{e^{\beta\overline{x}/2} + e^{-\beta\overline{x}/2}} - \overline{x} = \tanh(\tfrac{1}{2}\beta\overline{x}) - \overline{x}. \end{aligned} \tag{3.3}$$

Note that the constant N cancels out of this formula. When N is large, as long as the process is near the point $\overline{x}$, it locally behaves as a rescaled random walk with drift $g_\beta(\overline{x})$. In view of this, by some law of large numbers (that will be made rigorous in Theorem 3.2), we expect

$(\overline{X}_t)_{t\geq 0}$ to converge in distribution, as $N \to \infty$, to a solution of the differential equation

$$\tfrac{\partial}{\partial t}\overline{X}_t = g_\beta(\overline{X}_t) \qquad (t \geq 0). \tag{3.4}$$

3.3 Analysis of the Mean-Field Model

Assuming the correctness of (3.4) for the moment, we can study the behavior of the mean-field Ising model $\overline{X}$ in the limit that we first send $N \to \infty$, and then $t \to \infty$. A simple analysis of the function g_β (see Figure 3.2) reveals that the differential equation (3.4) has a single fixed point for $\beta \leq 2$ and three fixed points for $\beta > 2$. Here, with a *fixed point* of the differential equation, we mean a point z such that $\overline{x}_0 = z$ implies $\overline{x}_t = z$ for all $t \geq 0$; that is, this is a point such that $g_\beta(z) = 0$.

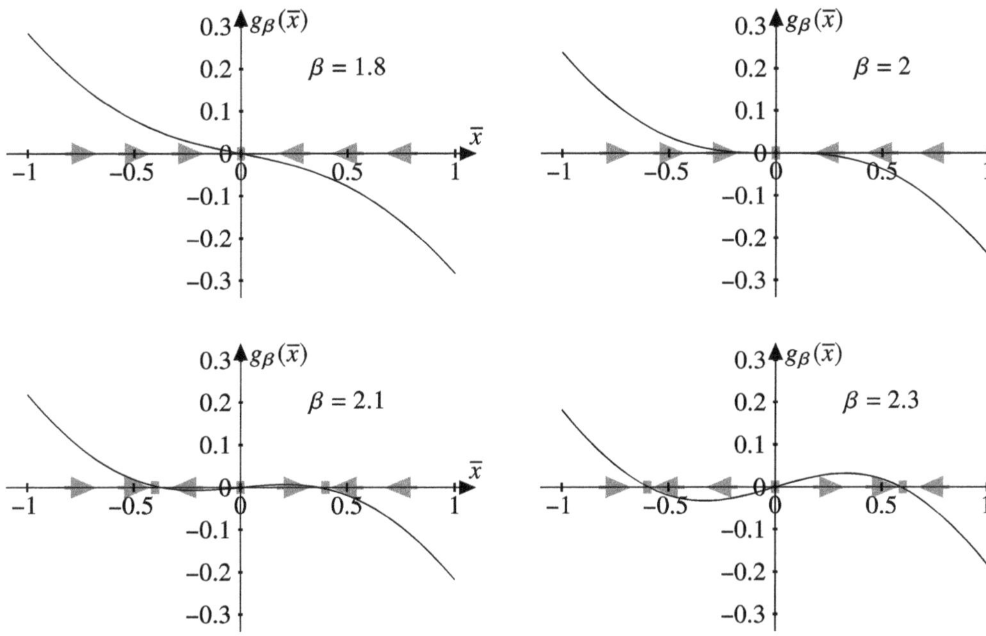

Figure 3.2 The drift function g_β for $\beta = 1.8$, 2, 2.1, and 2.3, respectively. For $\beta > 2$, the fixed point $\overline{x} = 0$ becomes unstable, and two new fixed points appear.

Indeed, using the facts that tanh is an odd function that is concave on $[0, \infty)$ and satisfies $\frac{\partial}{\partial x}\tanh(x)|_{x=0} = 1$, we see that:

- For $\beta \leq 2$, the equation $g_\beta(x) = 0$ has the unique solution $x = 0$.
- For $\beta > 2$, the equation $g_\beta(x) = 0$ has three solutions $x_- < 0 < x_+$.

For $\beta \leq 2$, solutions to the differential equation (3.4) converge to the unique fixed point $x = 0$ as time tends to infinity. On the other hand, for $\beta > 2$, the fixed point $x = 0$ becomes unstable. Solutions $\overline{X}$ to the differential equation (3.4) starting in $\overline{X}_0 > 0$ converge to x_+, while solutions starting in $\overline{X}_0 < 0$ converge to x_-.

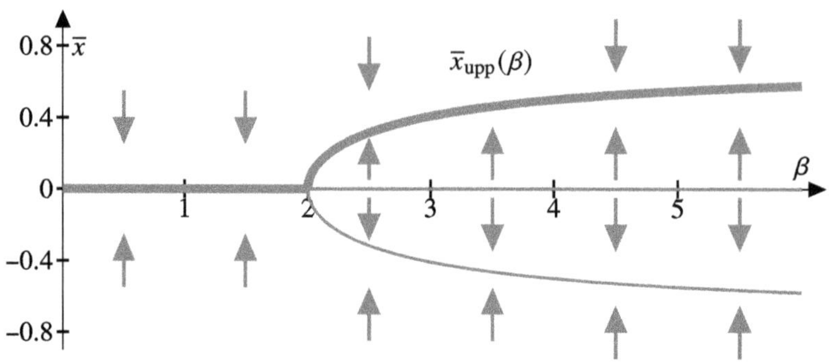

Figure 3.3 Fixed points of the mean-field Ising model as a function of β, with their domains of attraction. The upper fixed point as a function of β is indicated with a bold line.

In Figure 3.3, we have plotted the three fixed points $x_- < 0 < x_+$ as a function of β and indicated their domains of attraction. The function

$$x_{\text{upp}}(\beta) := \begin{cases} 0 & \text{if } \beta \leq 2, \\ \text{the unique positive solution of } \tanh(\tfrac{1}{2}\beta x) = x & \text{if } \beta > 2 \end{cases} \tag{3.5}$$

plays a similar role as the spontaneous magnetization $m_*(\beta)$ for the Ising model on $\mathbb{Z}^d$ (see formula (1.15)). More precisely, for mean-field processes started in initial states $\overline{X}_0 > 0$, the quantity x_{upp} describes the double limit

$$\lim_{t\to\infty} \lim_{N\to\infty} \overline{X}_t = x_{\text{upp}}. \tag{3.6}$$

We see from (3.5) that the mean-field Ising model (as formulated in (3.1)) exhibits a second-order (i.e., continuous) phase transition at the critical point $\beta_c = 2$. Since

$$x_{\text{upp}}(\beta) \propto (\beta - \beta_c)^{1/2} \quad \text{as} \quad \beta \downarrow \beta_c,$$

the *mean-field critical exponent* associated with the magnetization[1] is $c = 1/2$, which is the same as for the Ising model on $\mathbb{Z}^d$ in dimensions $d \geq 4$ (see Section 1.5). Understanding why the mean-field model correctly predicts the critical exponent in sufficiently high dimensions goes beyond the scope of the present chapter.

To conclude the present section, we note that the two limits in (3.6) cannot be interchanged. Indeed, for each fixed N, the Markov process $\overline{X}$ is irreducible, and hence, by Theorem 2.4, has a unique equilibrium law that is the long-time limit of the law at time t, started from an arbitrary initial state. In view of the symmetry of the problem, the magnetization in equilibrium must be zero, so regardless of the initial state, we have, for each fixed N,

$$\lim_{t\to\infty} \mathbb{E}[\overline{X}_t] = 0.$$

The reason why this can be true while at the same time (3.6) also holds is that the speed of convergence to equilibrium of the Markov process $\overline{X}$ becomes very slow as $N \to \infty$.

[1] In general, for a given second-order phase transition, there are several quantities of interest that all show power-law behavior near the critical point, and hence there are also several critical exponents associated with a given phase transition.

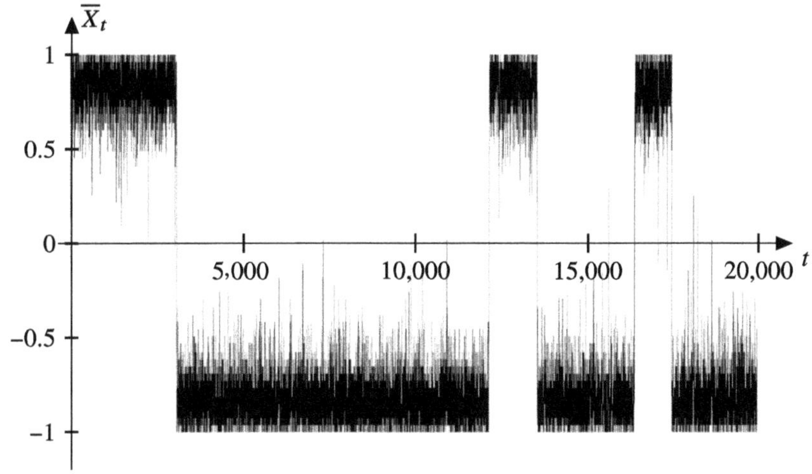

Figure 3.4 Metastable behavior of a mean-field Ising model with $N = 50$ and $\beta = 3$. Note the different time scale compared to Figure 3.1.

In Figure 3.4, we have plotted the time evolution of a mean-field Ising model $\overline{X}$ on a lattice with $N = 50$ sites, for a value of β above the critical point (concretely $\beta = 3$, which lies above $\beta_c = 2$). Although the average of $\overline{X}$ in the long run is 0, we see that the process spends most of its time around the values x_{upp} and $-x_{\text{upp}}$, with rare transitions between the two. This sort of behavior is called *metastable behavior.*

The value $N = 50$ was near the highest possible value for which I could still numerically observe this sort of behavior. For $N = 100$, the transitions between the two metastable states x_{upp} and $-x_{\text{upp}}$ become so rare that my program was no longer able to see them within a reasonable runtime. With the help of *large deviations theory*, one can show that the time that the system spends in one metastable state is approximately exponentially distributed (with a large mean) and calculate the asymptotics of the mean waiting time as $N \to \infty$. It turns out that the mean time one has to wait for a transition grows exponentially fast in N.

3.4 Functions of Markov Processes

In the present section, we formulate a proposition and a theorem that we have already implicitly used. Both are concerned with functions of Markov processes. Let $X = (X_t)_{t\geq 0}$ be a Markov process with finite state space S, generator G, and semigroup $(P_t)_{t\geq 0}$. Let T be another finite set and let $f\colon S \to T$ be a function. For each $x \in S$ and $y' \in T$ such that $f(x) \neq y'$, let

$$\mathcal{H}(x, y') := \sum_{x' \in S:\, f(x') = y'} G(x, x') \tag{3.7}$$

be the total rate at which $f(X_t)$ jumps to the state y', when the present state is $X_t = x$. The next proposition says that if these rates are a function of $f(x)$ only, then the process $Y = (Y_t)_{t\geq 0}$ defined by

$$Y_t := f(X_t) \qquad (t \geq 0) \tag{3.8}$$

is itself a Markov process.

Proposition 3.1 (Autonomous Markov process) *Assume that the rates in (3.7) are of the form*

$$\mathcal{H}(x,y') = H\big(f(x),y'\big) \qquad (x \in S,\ y' \in T,\ f(x) \neq y') \tag{3.9}$$

where H is a Markov generator of some process in T. Then the process Y defined in (3.8) is a Markov process with generator H. Conversely, if for each initial law of the process X, it is true that Y is a Markov process with generator H, then (3.9) must hold.

Proof of Proposition 3.1 Let us define $\mathcal{H}(x,y')$ as in (3.7) also when $f(x) = y'$. We start by noting that if (3.9) holds for all $x \in S$ and $y' \in T$ such that $f(x) \neq y'$, then it also holds when $f(x) = y'$. To see this, we write

$$\begin{aligned} H\big(f(x),f(x)\big) &= -\sum_{y':\,y'\neq f(x)} H(f(x),y') = -\sum_{y':\,y'\neq f(x)} \mathcal{H}(x,y') \\ &= -\sum_{y':\,y'\neq f(x)}\ \sum_{x':\,f(x')=y'} G(x,x') = -\sum_{x':\,f(x')\neq f(x)} G(x,x') = \sum_{x':\,f(x')=f(x)} G(x,x'), \end{aligned}$$

where we have used that since H and G are Markov generators, one has $\sum_{y'\in T} H(f(x),y') = 0$ and $\sum_{x'\in S} G(x,x') = 0$. We have thus shown that (3.9) is equivalent to

$$H\big(f(x),y'\big) = \sum_{x':\,f(x')=y'} G(x,x') \qquad \big(x \in S,\ y' \in T\big). \tag{3.10}$$

We claim that this is equivalent to

$$Q_t\big(f(x),y'\big) = \sum_{x':\,f(x')=y'} P_t(x,x') \qquad \big(t \geq 0,\ x \in S,\ y' \in T\big), \tag{3.11}$$

where $(Q_t)_{t\geq 0}$ is the semigroup generated by H. To prove this, we start by observing that for any function $g : T \to \mathbb{R}$,

$$\begin{aligned} G(g\circ f)(x) &= \sum_{x'} G(x,x')g(f(x')) = \sum_{y'}\ \sum_{x':\,f(x')=y'} G(x,x')g(y'), \\ (Hg)\circ f(x) &= \sum_{y'} H(f(x),y')g(y'). \end{aligned}$$

The right-hand sides of these equations are equal for all $g : T \to \mathbb{R}$ if and only if (3.10) holds, so (3.10) is equivalent to the statement that

$$G(g\circ f) = (Hg)\circ f \qquad (g : T \to \mathbb{R}). \tag{3.12}$$

By exactly the same argument with G replaced by P_t and H replaced by Q_t, we see that (3.11) is equivalent to

$$P_t(g\circ f) = (Q_t g)\circ f \qquad (t \geq 0,\ g : T \to \mathbb{R}). \tag{3.13}$$

To see that (3.12) and (3.13) are equivalent, we write

$$P_t = e^{Gt} = \sum_{n=0}^{\infty} \frac{1}{n!} t^n G^n \quad \text{and} \quad Q_t = e^{Ht} = \sum_{n=0}^{\infty} \frac{1}{n!} t^n H^n. \tag{3.14}$$

We observe that (3.12) implies

$$G^2(g \circ f) = G\big((Hg) \circ f\big) = (H^2 g) \circ f,$$

and similarly, by induction, $G^n(g \circ f) = (H^n g) \circ f$ for all $n \geq 0$, which by (3.14) implies (3.13). Conversely, if (3.13) holds for all $t \geq 0$, then it must hold up to first order in t as $t \downarrow 0$, which implies (3.12). This completes the proof that (3.9) is equivalent to (3.11).

If (3.11) holds, then by (2.6), the finite-dimensional distributions of Y are given by

$$\begin{aligned} &\mathbb{P}\big[Y_0 = y_0, \ldots, Y_{t_n} = y_n\big] \\ &= \sum_{x_0:\, f(x_0)=y_0} \cdots \sum_{x_n:\, f(x_n)=y_n} \mathbb{P}[X_0 = x_0] P_{t_1 - t_0}(x_0, x_1) \cdots P_{t_n - t_{n-1}}(x_{n-1}, x_n) \\ &= \mathbb{P}[Y_0 = y_0] Q_{t_1 - t_0}(y_0, y_1) \cdots Q_{t_n - t_{n-1}}(y_{n-1}, y_n) \end{aligned} \tag{3.15}$$

$(0 = t_0 < \cdots < t_n)$. Again by (2.6), this implies that Y is a Markov process with generator H.

Conversely, if Y is a Markov process with generator H for each initial state of X, then for each $x_0 \in S$, (3.15) must hold when $X_0 = x_0$ a.s. and for $n = 1$, from which we see that (3.11) and hence (3.9) hold. ∎

Summarizing, Proposition 3.1 says that if $Y_t = f(X_t)$ is a function of a Markov process, and the jump rates of Y are a function of the present state of Y only (and do not otherwise depend on the state of X), then Y is itself a Markov process. In such a situation, we will say that Y is an *autonomous* Markov process. We have already implicitly used Proposition 3.1 in Section 3.2, when we claimed that the process $\overline{X}$ is a Markov process with jump rates as in (3.2).

Remark For the final statement of the proposition, it is essential that Y is a Markov process for *each* initial law X. There exist interesting examples of functions of Markov processes that are not autonomous Markov processes but nonetheless are Markov processes for some *special* initial laws of the original Markov process. This is closely related to the concept of intertwining of Markov processes that will briefly be mentioned in Section 6.5.

Our next aim is to make the claim rigorous that for large N, the process $\overline{X}$ can be approximated by solutions to the differential equation (3.4). The basic idea is that since the process makes many small steps, as long as $\overline{X}_t \approx \overline{x}$, by some sort of "local" law of large numbers, the process should deterministically increase at speed $\approx g_\beta(\overline{x})$. To make this precise, we will apply a theorem from [DN08]. Although the proof is not very complicated, it is a bit lengthy and would detract from our main objects of interest here, so we only show how the theorem below can be deduced from a theorem in [DN08]. That paper also treats the multidimensional case and gives explicit estimates on probabilities of the form (3.19) below. An alternative, more probabilistic approach to mean-field equations is described in [MSS20].

For each $N \geq 1$, let $X^N = (X^N_t)_{t \geq 0}$ be a Markov process with finite state space S_N, generator G_N, and semigroup $(P^N_t)_{t \geq 0}$, and let $f_N : S_N \to \mathbb{R}$ be functions. We will be interested in conditions under which the processes $(f_N(X^N_t))_{t \geq 0}$ approximate the solution $(y_t)_{t \geq 0}$ of a differential equation, in the limit $N \to \infty$. Note that we do not require that $f_N(X^N_t)$ is an autonomous Markov process. To ease notation, we will sometimes drop the super- and subscripts N when no confusion arises.

We define two functions $\alpha = \alpha_N$ and $\beta = \beta_N$ that describe the quadratic variation and drift, respectively, of the process $f(X_t)$. More precisely, these functions are given by

$$\alpha(x) := \sum_{x' \in S} G(x, x')\big(f(x') - f(x)\big)^2,$$
$$\beta(x) := \sum_{x' \in S} G(x, x')\big(f(x') - f(x)\big).$$

The idea is that if α tends to zero and β approximates a nice, Lipschitz continuous function of $f(X_t)$, then $f(X_t)$ should in the limit be given by the solution of a differential equation.

We assume that the functions f_N all take values in a closed interval $I \subset \mathbb{R}$ with left and right boundaries $I_- := \inf I$ and $I_+ := \sup I$, which may be finite or infinite. We also assume that there exists a globally Lipschitz function $b\colon I \to \mathbb{R}$ such that

$$\sup_{x \in S_N} \big|\beta_N(x) - b\big(f_N(x)\big)\big| \underset{N\to\infty}{\longrightarrow} 0, \tag{3.16}$$

that is, the drift function β is uniformly approximated by $b \circ f_N$. Assuming also that

$$b(I_-) \geq 0 \quad \text{if } I_- > -\infty \quad \text{and} \quad b(I_+) \leq 0 \quad \text{if } I_+ < -\infty, \tag{3.17}$$

the differential equation

$$\tfrac{\partial}{\partial t} y_t = b(y_t) \qquad (t \geq 0)$$

has a unique I-valued solution $(y_t)_{t\geq 0}$ for each initial state $y_0 \in I$. The following theorem gives sufficient conditions for the I-valued processes $(f_N(X_t^N))_{t\geq 0}$ to approximate a solution of the differential equation.

Theorem 3.2 (Limiting differential equation) *Assume that $f_N(X_0^N)$ converges in probability to y_0 and that as well as (3.16), one moreover has*

$$\sup_{x \in S_N} \alpha_N(x) \underset{N\to\infty}{\longrightarrow} 0. \tag{3.18}$$

Then, for each $T < \infty$ and $\varepsilon > 0$,

$$\mathbb{P}\big[|f_N(X_t^N) - y_t| \leq \varepsilon \ \forall t \in [0, T]\big] \underset{N\to\infty}{\longrightarrow} 1. \tag{3.19}$$

Proof We apply [DN08, Thm 4.1]. Fix $T < \infty$ and $\varepsilon > 0$, and also fix $y_0 \in I$. Let L denote the Lipschitz constant of b. The assumptions of [DN08, Thm 4.1] allow for the case that f_N does not in general take values in I, but only under the additional condition that $f_N(x)$ is not further than ε from a possible value of the solution of the differential equation. In our case, these more general assumptions are automatically satisfied. Set $\delta := \frac{1}{3}\varepsilon e^{-LT}$. We consider the events

$$\Omega_0 := \big\{|f(X_0) - y_0| \leq \delta\big\} \quad \text{and} \quad \Omega_1 := \left\{\int_0^T |\beta(X_t) - b\big(f(X_t)\big)|\,\mathrm{d}t \leq \delta\right\}.$$

For $K > 0$, we also define

$$\Omega_{K,2} := \left\{\int_0^T \alpha(X_t)\,\mathrm{d}t \leq KT\right\}.$$

Then [DN08, Thm 4.1] tells us that

$$\mathbb{P}\big[\sup_{t\in[0,T]}|f(X_t)-y_t|>\varepsilon\big]\leq 4KT\delta^{-2}+\mathbb{P}\big(\Omega_0^{\mathrm{c}}\cup\Omega_1^{\mathrm{c}}\cup\Omega_{K,2}^{\mathrm{c}}\big). \tag{3.20}$$

Our assumption that $f_N(X_0^N)\to y_0$ in probability implies that $\mathbb{P}(\Omega_0^{\mathrm{c}})\to 0$ as $N\to\infty$. Set

$$A_N:=\sup_{x\in S_N}\alpha_N(x)\quad\text{and}\quad B_N:=\sup_{x\in S_N}\big|\beta_N(x)-b\big(f_N(x)\big)\big|$$

Then $A_N\to 0$ by (3.18) and $B_N\to 0$ by (3.16). Since

$$\int_0^T|\beta(X_t)-b\big(f(X_t)\big)|\,\mathrm{d}t\leq B_NT\leq\delta$$

for N sufficiently large, we see that $\mathbb{P}(\Omega_1^{\mathrm{c}})=0$ for N sufficiently large. Also, since

$$\int_0^T\alpha(X_t)\,\mathrm{d}t\leq A_NT,$$

we see that $\mathbb{P}(\Omega_{A_N,2}^{\mathrm{c}})=0$ for all N. Inserting $K=A_N$ in (3.20), we see that the right-hand side tends to zero as $N\to\infty$. ∎

Using Theorem 3.2, we can make the approximation of the mean-field Ising model by the differential equation (3.4) rigorous. Let $X^N=(X_t^N)_{t\geq 0}$ denote the Markov process with state space $\{-1,+1\}^{\Lambda_N}$, where Λ_N is a set containing N elements and the jump rates of X^N are given in (3.1). By Proposition 3.1, the process $\overline{X}_t^N:=\frac{1}{N}\sum_{i\in\Lambda_N}X_t(i)$ is itself a Markov process with jump rates as in (3.2). We can either apply Theorem 3.2 directly to the Markov processes X^N and the functions $f_N(x):=\frac{1}{N}\sum_{i\in\Lambda_N}x(i)$, or we can apply Theorem 3.2 to the Markov processes $\overline{X}^N$ and choose for f_N the identity function $f_N(\overline{x})=\overline{x}$. In either case, the assumption (3.16) has already been verified in (3.3). To check also (3.18), we calculate

$$\alpha_N(x)=r_+^N(\overline{x})\Big(\frac{2}{N}\Big)^2+r_-^N(\overline{x})\Big(\frac{2}{N}\Big)^2=\frac{2}{N}\Big(1+\overline{x}\frac{e^{-\beta\overline{x}/2}-e^{\beta\overline{x}/2}}{e^{-\beta\overline{x}/2}+e^{\beta\overline{x}/2}}\Big),$$

which clearly tends uniformly to zero as $N\to\infty$.

3.5 The Mean-Field Contact Process

Recall the definition of the generator of the contact process from (1.8). We slightly reformulate this as

$$\begin{aligned}G_{\mathrm{cont}}f(x):=&\lambda\sum_{i\in\mathbb{Z}^d}\frac{1}{|\mathcal{N}_i|}\sum_{j\in\mathcal{N}_i}\big\{f\big((\mathtt{bra}_{ij}(x)\big)-f\big(x\big)\big\}\\&+\sum_{i\in\mathbb{Z}^d}\big\{f\big((\mathtt{death}_i(x)\big)-f\big(x\big)\big\}\qquad(x\in\{0,1\}^\Lambda),\end{aligned}\tag{3.21}$$

where as customary, we have set the death rate to $\delta=1$, and we have also reparametrized the infection rate so that λ denotes the total rate of all outgoing infections from a given site, instead of the infection rate per neighbor.

We will be interested in the contact process on the complete graph, which means that we take for $\Lambda=\Lambda_N$ a set with N elements, which we equip with the structure of a complete graph with (undirected) edge set $E=E_N:=\{\{i,j\}:i,j\in\Lambda_N\}$ and corresponding set of directed edges $\mathcal{E} = \mathcal{E}_N$. We view i as a neighbor of itself, but since $\mathtt{bra}_{ii}$ is the identity map, this has no effect. We will be interested in the fraction of infected sites

$$\overline{X}_t = \overline{X}_t^N := \frac{1}{N} \sum_{i \in \Lambda_N} X_t(i) \qquad (t \geq 0),$$

which jumps with the following rates

$$\begin{aligned} \overline{x} &\mapsto \overline{x} + \tfrac{1}{N} \qquad \text{with rate} \qquad r_+^N(\overline{x}) := \lambda N \overline{x}(1 - \overline{x}), \\ \overline{x} &\mapsto \overline{x} - \tfrac{1}{N} \qquad \text{with rate} \qquad r_-^N(\overline{x}) := N \overline{x}. \end{aligned} \tag{3.22}$$

Here $N(1 - \overline{x})$ is the number of healthy sites, each of which gets infected with rate $\lambda \overline{x}$, and $N\overline{x}$ is the number of infected sites, each of which recovers with rate one. Note that since these rates are a function of $\overline{x}$ only, by Proposition 3.1, the process $(\overline{X}_t)_{t \geq 0}$ is an autonomous Markov process.

We wish to apply Theorem 3.2 to conclude that $\overline{X}$ can for large N be approximated by the solution of a differential equation. To this aim, we calculate the drift β and the quadratic variation function α.

$$\begin{aligned} \alpha_N(x) &= r_+^N(\overline{x}) \tfrac{1}{N^2} + r_-^N(\overline{x}) \tfrac{1}{N^2} = \frac{1}{N} \big(\lambda \overline{x}(1 - \overline{x}) + \overline{x} \big), \\ \beta_N(x) &= r_+^N(\overline{x}) \tfrac{1}{N} - r_-^N(\overline{x}) \tfrac{1}{N} = \lambda \overline{x}(1 - \overline{x}) - \overline{x}. \end{aligned}$$

By Theorem 3.2, it follows that in the mean-field limit $N \to \infty$, the fraction of infected sites can be approximated by solutions of the differential equation

$$\tfrac{\partial}{\partial t} \overline{X}_t = b_\lambda(\overline{X}_t) \quad (t \geq 0), \quad \text{where} \quad b_\lambda(\overline{x}) := \lambda \overline{x}(1 - \overline{x}) - \overline{x}. \tag{3.23}$$

The equation $b_\lambda(\overline{x}) = 0$ has the solutions

$$\overline{x} = 0 \quad \text{and} \quad \overline{x} = 1 - \lambda^{-1}.$$

The second solution lies inside the interval $[0, 1]$ of possible values of $\overline{X}_t$ if and only if $\lambda \geq 1$. Plotting the function b_λ for $\lambda < 1$ and $\lambda > 1$ yields Figure 3.5.

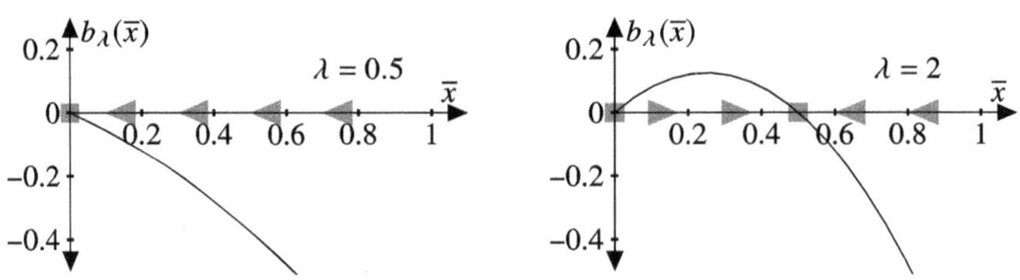

Figure 3.5 The function b_λ for two values of λ.

We see from this that the fixed point $\overline{x} = 0$ is stable for $\lambda \leq 1$ but becomes unstable for $\lambda > 1$, in which case $\overline{x} = 1 - \lambda^{-1}$ is the only stable fixed point that attracts all solutions started in a nonzero initial state. The situation is summarized in Figure 3.6.

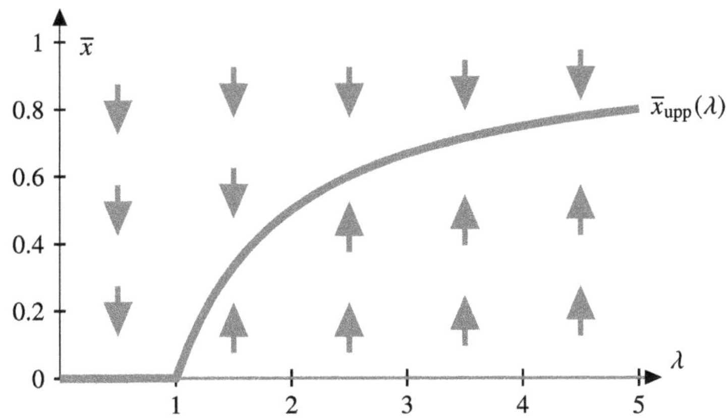

Figure 3.6 Mean-field analysis of the contact process.

Letting $\overline{x}_{\rm upp}(\lambda) := 0 \vee (1 - \lambda^{-1})$ denote the stable fixed point, we see that the mean-field contact process exhibits a second-order phase transition at the critical point $\lambda_{\rm c} = 1$. Since

$$\overline{x}_{\rm upp}(\lambda) \propto (\lambda - \lambda_{\rm c}) \quad \text{as } \lambda \downarrow \lambda_{\rm c},$$

the associated critical exponent is $c = 1$, in line with what we know for contact processes in dimensions $d \geq 4$ (see the discussion in Section 1.5).

3.6 The Mean-Field Voter Model

Recall the definition of the generator of the voter model from (1.5). For simplicity, we will only consider the two-type model and as the local state space we will choose $S = \{0, 1\}$. Specializing to the complete graph $\Lambda = \Lambda_N$ with N vertices, the generator becomes

$$G_{\rm vot} f(x) = \frac{1}{|\Lambda|} \sum_{(i,j)\in\mathcal{E}} \{f((\mathtt{vot}_{ij}(x)) - f(x)\} \qquad (x \in \{0,1\}^{\Lambda}). \tag{3.24}$$

Note that the factor $|\Lambda|^{-1}$ says that each site i updates its type with rate one, and at such an event chooses a new type from a uniformly chosen site j (allowing for the case $i = j$, which has no effect).

We are interested in the fraction of sites of type 1,

$$\overline{X}_t = \overline{X}^N_t := \frac{1}{N} \sum_{i\in\Lambda_N} X_t(i) \qquad (t \geq 0),$$

which jumps as (compare (3.22))

$$\begin{aligned} \overline{x} &\mapsto \overline{x} + \tfrac{1}{N} \quad &&\text{with rate} \quad r^N_+(\overline{x}) := N\overline{x}(1 - \overline{x}), \\ \overline{x} &\mapsto \overline{x} - \tfrac{1}{N} \quad &&\text{with rate} \quad r^N_-(\overline{x}) := N\overline{x}(1 - \overline{x}). \end{aligned}$$

Note that $N(1 - \overline{x})$ is the number of sites of type 0, and that each such site adopts the type

1 with rate $\overline{x}$. The derivation of $r_-^N(\overline{x})$ is similar. We calculate the drift β and quadratic variation function α.

$$\alpha_N(x) = r_+^N(\overline{x})\tfrac{1}{N^2} + r_-^N(\overline{x})\tfrac{1}{N^2} = \frac{2}{N}\overline{x}(1-\overline{x}),$$
$$\beta_N(x) = r_+^N(\overline{x})\tfrac{1}{N} - r_-^N(\overline{x})\tfrac{1}{N} = 0.$$

Applying Theorem 3.2, we see that in the limit $N \longrightarrow \infty$, the process $(\overline{X}_t)_{t\geq 0}$ is well approximated by solutions to the differential equation

$$\tfrac{\partial}{\partial t}\overline{X}_t = 0 \qquad (t \geq 0),$$

that is, $\overline{X}_t$ is approximately constant as a function of t.

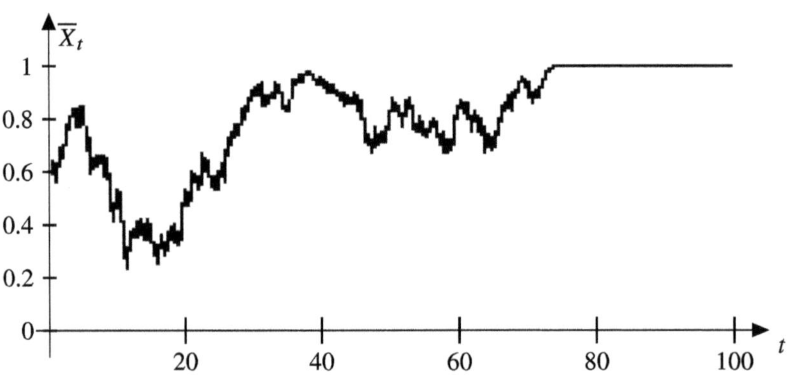

Figure 3.7 The fraction of type 1 individuals in the mean-field voter model from (3.25) on $N = 100$ sites. This process approximates the Wright–Fisher diffusion.

Of course, if we go to larger time scales, then $\overline{X}_t$ will no longer be constant; compare Figure 3.4. In fact, we can determine the time scale at which $\overline{X}_t$ fluctuates quite precisely. Speeding up time by a factor $|\Lambda| = N$ is the same as multiplying all rates by a factor $|\Lambda|$. If we repeat our previous calculations for the process with generator

$$G_{\text{vot}}f(x) = \sum_{(i,j)\in\mathcal{E}} \big\{f\big((\texttt{vot}_{ij}(x)\big) - f\big(x\big)\big\} \qquad (x \in \{0,1\}^{\Lambda}), \tag{3.25}$$

then the drift and quadratic variation are given by

$$\alpha_N(x) = 2\overline{x}(1-\overline{x}),$$
$$\beta_N(x) = 0.$$

In this case, the quadratic variation does not go to zero, so Theorem 3.2 is no longer applicable. One can show, however, that in the limit $N \longrightarrow \infty$, the new, sped-up process is well approximated by solutions to the (Itô) stochastic differential equation (SDE)

$$\mathrm{d}\overline{X}_t = \sqrt{2\overline{X}_t(1-\overline{X}_t)}\,\mathrm{d}B_t \qquad (t \geq 0),$$

where $2\overline{X}_t(1-\overline{X}_t) = \alpha(X_t)$ is, of course, the quadratic variation function we have just calculated. Solutions to this SDE are *Wright–Fisher diffusions*, that is, Markov processes with continuous sample paths and generator

$$Gf(\overline{x}) = \overline{x}(1-\overline{x})\tfrac{\partial^2}{\partial \overline{x}^2} f(\overline{x}). \tag{3.26}$$

These calculations can be made rigorous using methods from the theory of convergence of Markov processes; see, for example, the book [EK86]. See Figure 3.7 for a simulation of the process $\overline{X}$ when X has the generator in (3.25) and $N = 100$.

3.7 Exercises

Exercise 3.3 *Do a mean-field analysis of the process with generator*

$$Gf(x) = b|\Lambda|^{-2}\sum_{ii'j}\{f(\mathtt{coop}_{ii'j}(x)) - f(x)\} + \sum_{i}\{f(\mathtt{death}_i(x)) - f(x)\},$$

where the maps $\mathtt{coop}_{ii'j}$ *and* $\mathtt{death}_i$ *are defined in (1.27) and (1.7), respectively. Do you observe a phase transition? Is it first or second order?* Hint: *Figure 3.8.*

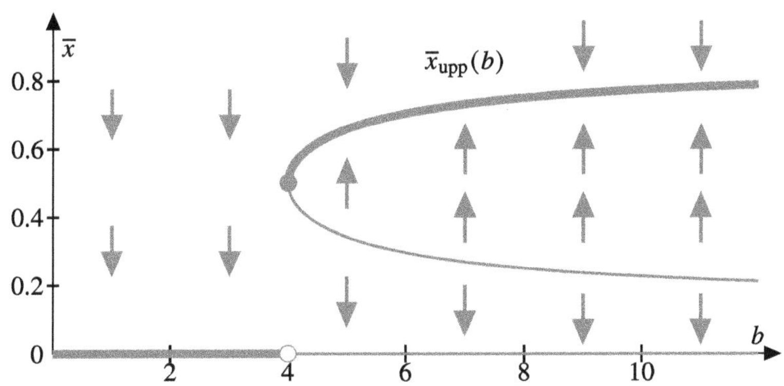

Figure 3.8 Mean-field analysis of a model with cooperative branching and deaths.

Exercise 3.4 *Same as above for the model with generator*

$$Gf(x) = b|\Lambda|^{-2}\sum_{ii'j}\{f(\mathtt{coop}_{ii'j}(x)) - f(x)\} + |\Lambda|^{-1}\sum_{ij}\{f(\mathtt{rw}_{ij}(x)) - f(x)\}.$$

Exercise 3.5 *Derive an SDE in the limit* $|\Lambda| \to \infty$ *for the density of the mean-field voter model with small bias and death rates, with generator*

$$\begin{aligned} Gf(x) = &\sum_{ij\in\Lambda}\{f(\mathtt{vot}_{ij}(x)) - f(x)\} + s|\Lambda|^{-1}\sum_{ij\in\Lambda}\{f(\mathtt{bra}_{ij}(x)) - f(x)\} \\ &+ d\sum_{i\in\Lambda}\{f(\mathtt{death}_i(x)) - f(x)\}. \end{aligned}$$

Hint: *You should find expressions of the form*

$$\begin{aligned} \mathbb{E}^{\overline{x}}\big[(\overline{X}_t - \overline{x})\big] &= b(\overline{x})\cdot t + O(t^2), \\ \mathbb{E}^{\overline{x}}\big[(\overline{X}_t - \overline{x})^2\big] &= a(\overline{x})\cdot t + O(t^2), \end{aligned}$$

which leads to a limiting generator of the form

$$Gf(\overline{x}) = \tfrac{1}{2}a(\overline{x})\tfrac{\partial^2}{\partial \overline{x}^2} f(\overline{x}) + b(\overline{x})\tfrac{\partial}{\partial x} f(\overline{x}).$$

Exercise 3.6 *Do a mean-field analysis of the following more general version of the Neuhauser–Pacala model [NP99] from (1.18). In the more general model, the site i flips*

$$\begin{aligned} 0 \mapsto 1 \quad &\text{with rate} \quad \big(f_0 + \alpha_{01} f_1\big) f_1, \\ 1 \mapsto 0 \quad &\text{with rate} \quad \big(f_1 + \alpha_{10} f_0\big) f_0, \end{aligned}$$

where $\alpha_{01}, \alpha_{10} > 0$ *and* $f_\tau = |\mathcal{N}_i|^{-1} \sum_{j \in \mathcal{N}_i} 1_{\{x(j)=\tau\}}$ *is the relative frequency of type* τ *in the neighborhood of i. Find all stable and unstable fixed points of the mean-field model in the regimes: I.* $\alpha_{01}, \alpha_{10} < 1$, *II.* $\alpha_{01} < 1 < \alpha_{10}$, *III.* $\alpha_{10} < 1 < \alpha_{01}$, *IV.* $1 < \alpha_{01}, \alpha_{10}$.

Exercise 3.7 *Do a mean-field analysis of the cycle conform model introduced in Section 1.9.* Warning: *this is considerably more complicated than the previous exercises. Working hard enough, it should be possible to establish the following facts. Let* $F_t(s)$ *denote the fraction of vertices that at time t are in the local state* $s \in \{0, 1, 2\}$. *Then, in the mean-field limit, one has*

$$F_t(s) = \tfrac{1}{3} + \tfrac{2}{3}\Re(e^{is\phi} f_t) \qquad (s \in \{0,1,2\},\ t \geq 0),$$

where $\phi := 2\pi/3$ *and* $t \mapsto f_t \in \mathbb{C}$ *solves the differential equation*

$$\tfrac{\partial}{\partial t} f_t = \tfrac{1}{3}\alpha\big\{f_t + f_t^* f_t^* - 2 f_t f_t f_t^*\big\} + (1-\alpha)\big(e^{-i\phi} - 1\big) f_t \qquad (t \geq 0).$$

Here $\Re(z)$ *denotes the real part of a complex number z and* z^* *denotes its complex conjugate. For* $0 \leq \alpha < 6/7$, *the only fixed point of this equation is* $f_t = 0$. *This fixed point is stable for* $\alpha < 9/11$ *but unstable for* $\alpha > 9/11$. *In the intermediate regime* $9/11 < \alpha < 6/7$, *there are periodic solutions.*

4

Construction and Ergodicity

4.1 Introduction

Throughout this chapter, S is a finite set called the *local state space*, Λ is a countable set called the *lattice*, $\mathcal{G}$ is a countable collection of continuous maps $m\colon S^\Lambda \to S^\Lambda$, and $(r_m)_{m\in\mathcal{G}}$ are nonnegative rates. We aim to construct a Markov process with state space S^Λ and formal generator of the form

$$Gf(x) = \sum_{m\in\mathcal{G}} r_m\{f(m(x)) - f(x)\} \qquad (x \in S^\Lambda). \tag{4.1}$$

We will usually be interested in the case that all $m \in \mathcal{G}$ are local maps, that is, the set $\mathcal{D}(m)$ of lattice points whose values can be changed by m as defined in Section 2.8 is finite, but our proofs do not need the finiteness of $\mathcal{D}(m)$. Under the assumption

$$\sum_{m\in\mathcal{G}} r_m 1_{\mathcal{D}(m)}(i) < \infty \quad (i \in \Lambda),$$

one can show that Gf is well-defined for all functions $f\colon S^\Lambda \to \mathbb{R}$ that depend on finitely many coordinates. In general, we will need stronger conditions on the rates $(r_m)_{m\in\mathcal{G}}$ to ensure that G generates a Markov process. Most of the interacting particle systems introduced in Chapter 1 have a generator that can naturally be written in the form (4.1). The only processes for which this is not so obvious are the stochastic Ising and Potts models. Nevertheless, we will show in Section 4.7 below that also the generator of the stochastic Ising model with Glauber dynamics can be written in the form (4.1), and we will use this random mapping representation of G to prove ergodicity for small values of β.

The space S^Λ is uncountable except in the trivial case that S has cardinality one. This means that we cannot use the theory of continuous-time Markov chains. Instead, we will rely on the theory of Feller processes. In Section 4.2, we will collect some general facts about Feller processes, which are a class of Markov processes with compact, metrizable state spaces that are uniquely characterized by their generators. Since this is rather functional analytic material, which is moreover well-known, we will state the main facts without proof but give references to places where proofs can be found.

In Section 4.3, we then present a Poisson construction of interacting particle systems (including proofs) that is similar to the Poisson construction of continuous-time Markov chains. To some degree, this provides a probabilistic alternative to the functional analytic approach via Feller processes. To get the full picture, however, one needs both approaches, so in Section 4.5, we link the Feller process we have constructed in Section 4.3 to the generator defined in (4.1).

Luckily, all this abstract theory gives us more than just the information that the systems we are interested in are well defined. In Section 4.6, we will see that as a side result of our proofs, we can derive sufficient conditions for an interacting particle system to be ergodic, that is, to have a unique invariant law that is the long-time limit starting from any initial state.[1] In Section 4.7, we apply this to derive lower bounds on the critical points of the Ising model. The methods developed in Section 4.3 will also prove useful in Chapter 6, where we discuss duality.

4.2 Feller Processes

In Section 2.1, we showed how the semigroup of a Markov process on a finite state space can be characterized in terms of its generator, and in Section 2.3, we generalized this to countable state spaces. In the present section, we will treat a class of Markov processes with compact metrizable state spaces. The basic assumption we will make is that the transition probabilities $(P_t)_{t\geq 0}$ are continuous, which means that we will be discussing *Feller processes*. We will later apply the theory of Feller processes to state spaces of the form S^Λ equipped with the product topology, which are compact by Tychonoff's theorem. It is easy to see that the product topology on S^Λ is metrizable. For example, if $(a_i)_{i\in\Lambda}$ are strictly positive constants such that $\sum_{i\in\Lambda} a_i < \infty$, then

$$d(x,y) := \sum_{i\in\Lambda} a_i 1_{\{x(i)\neq y(i)\}}$$

defines a metric that generates the product topology.

Let E be a compact metrizable space.[2] We use the notation

$$\begin{aligned}\mathcal{B}(E) &:= \text{the Borel } \sigma\text{-field on } E,\\ B(E) &:= \text{the space of bounded, Borel-measurable functions } f\colon E\to\mathbb{R},\\ C(E) &:= \text{the space of continuous functions } f\colon E\to\mathbb{R},\\ \mathcal{M}_1(E) &:= \text{the space of probability measures } \mu \text{ on } E.\end{aligned}$$

We equip $C(E)$ with the supremum norm

$$\|f\|_\infty := \sup_{x\in E}|f(x)| \qquad (f\in C(E)),$$

making $C(E)$ into a Banach space. We equip $\mathcal{M}_1(E)$ with the topology of weak convergence, where by definition,[3] μ_n converges weakly to μ, denoted $\mu_n \Rightarrow \mu$, if $\int f\,\mathrm{d}\mu_n \to \int f\,\mathrm{d}\mu$ for all $f\in C(E)$. With this topology, $\mathcal{M}_1(E)$ is a compact metrizable space. The following lemma is often convenient.

Lemma 4.1 (Convergence criterion) *For $f_n, f \in C(E)$, the following conditions are equivalent:*

1 We use the word ergodic in a different meaning than in ergodic theory; see the discussion on page 96 in Section 4.6.

2 Such spaces are always separable and complete in any metric that generates the topology; in particular, they are Polish spaces.

3 More precisely, the topology of weak convergence is the unique *metrizable* topology with this property. Since in metrizable spaces, convergent subsequences uniquely characterize the topology, our definition is unambiguous.

(i) $\|f_n - f\|_\infty \underset{n\to\infty}{\longrightarrow} 0$,
(ii) $f_n(x_n) \underset{n\to\infty}{\longrightarrow} f(x)$ *for all* $x_n, x \in E$ *with* $x_n \to x$.

Proof Since

$$\big|f_n(x_n) - f(x)\big| \le \|f_n - f\|_\infty + \big|f(x_n) - f(x)\big|,$$

(i) and the continuity of f imply (ii). On the other hand, if (i) does not hold, then we can choose $x_n \in E$ such that $\limsup_{n\to\infty} |f_n(x_n) - f(x_n)| > 0$. By the compactness of E, going to a subsequence if necessary, we can assume that $x_n \to x$ for some $x \in E$. Since

$$\big|f_n(x_n) - f(x)\big| \ge \big|f_n(x_n) - f(x_n)\big| - \big|f(x_n) - f(x)\big|,$$

using the continuity of f, we see that $f_n(x_n) \not\to f(x)$. ∎

A *probability kernel* on E is a function $K\colon E \times \mathcal{B}(E) \to \mathbb{R}$ such that

(i) $K(x, \cdot\,)$ is a probability measure on E for each $x \in E$,
(ii) $K(\,\cdot\,, A)$ is a real measurable function on E for each $A \in \mathcal{B}(E)$.

This is equivalent to the statement that $x \mapsto K(x, \cdot\,)$ is a measurable map from E to $\mathcal{M}_1(E)$ (where the latter is equipped with the topology of weak convergence and the associated Borel σ-field). By definition, a probability kernel is *continuous* if the map $x \mapsto K(x, \cdot\,)$ is continuous (with respect to the topologies with which we have equipped these spaces). A probability kernel is *deterministic* if it is of the form $K(x, \cdot\,) = \delta_{m(x)}$ for some measurable map $m\colon E \to E$, where $\delta_{m(x)}$ denotes the delta-measure at $m(x)$. It is easy to see that a deterministic kernel is continuous if and only if m is a continuous map. A *random mapping representation* of a probability kernel K is a random measurable map[4] $M\colon E \to E$ such that $K(x, \cdot\,) = \mathbb{P}[M(x) \in \cdot\,]$ $(x \in E)$.[5]

If $K(x, \mathrm{d}y)$ is a probability kernel on a Polish space E, then setting

$$Kf(x) := \int_E K(x, \mathrm{d}y) f(y) \qquad \big(x \in E,\ f \in B(E)\big)$$

defines a linear operator $K\colon B(E) \to B(E)$. We define the *composition* of two probability kernels K, L as

$$(KL)(x, A) := \int_E K(x, \mathrm{d}y) L(y, A) \qquad \big(x \in E,\ A \in \mathcal{B}(E)\big).$$

Then KL is again a probability kernel on E, and the linear operator $(KL)\colon B(E) \to B(E)$ associated with this kernel is the composition of the linear operators K and L. It follows from the definition of weak convergence that a kernel K is continuous if and only if its associated linear operator maps the space $C(E)$ into itself. If μ is a probability measure and K is a probability kernel, then

$$(\mu K)(A) := \int \mu(\mathrm{d}x) K(x, A) \qquad \big(A \in \mathcal{B}(E)\big)$$

[4] More formally, this means that $M : \Omega \times E \to E$ is measurable with respect to the product-σ-field $\mathcal{F} \otimes \mathcal{B}(E)$, where $(\Omega, \mathcal{F}, \mathbb{P})$ is the underlying probability space.

[5] For infinite spaces, it is not so clear if every probability kernel has a random mapping representation. One could also ask if every continuous probability kernel has a representation in terms of continuous maps. Although these questions are interesting, we will neglect them here.

defines another probability measure μK. Introducing the notation $\mu f := \int f \, d\mu$, one has $(\mu K)f = \mu(Kf)$ for all $f \in B(E)$.

By definition, a *continuous transition probability* on E is a collection $(P_t)_{t \geq 0}$ of probability kernels on E, such that

(i) $(x,t) \mapsto P_t(x,\,\cdot\,)$ is a continuous map from $E \times [0,\infty)$ into $\mathcal{M}_1(E)$,
(ii) $P_0 = 1$ and $P_s P_t = P_{s+t} \quad (s,t \geq 0)$.

In particular, (i) implies that each P_t is a continuous probability kernel, so each P_t maps the space $C(E)$ into itself. One has

$$\begin{array}{ll} \text{(i)} & \lim_{t \to 0} P_t f = P_0 f = f \qquad (f \in C(E)), \\ \text{(ii)} & P_s P_t f = P_{s+t} f \qquad (s,t \geq 0), \\ \text{(iii)} & f \geq 0 \text{ implies } P_t f \geq 0, \\ \text{(iv)} & P_t 1 = 1, \end{array}$$

and conversely, each collection of linear operators $P_t \colon C(E) \to C(E)$ with these properties corresponds to a unique continuous transition probability on E. Such a collection of linear operators $P_t \colon C(E) \to C(E)$ is called a *Feller semigroup*. We note that in (i), the limit is (of course) with respect to the topology we have chosen on $C(E)$, that is, with respect to the supremum norm.

By definition, a function $w \colon [0,\infty) \to E$ is *cadlag* if it is right-continuous with left limits,[6] that is,

$$\begin{array}{lll} \text{(i)} & \lim_{t \downarrow s} w_t = w_s & (s \geq 0), \\ \text{(ii)} & \lim_{t \uparrow s} w_t =: w_{s-} \text{ exists} & (s > 0). \end{array}$$

Let $(P_t)_{t \geq 0}$ be a Feller semigroup. By definition a *Feller process* with semigroup $(P_t)_{t \geq 0}$ is a stochastic process $X = (X_t)_{t \geq 0}$ with cadlag sample paths[7] such that

$$\mathbb{P}\big[X_u \in \cdot \,\big|\, (X_s)_{0 \leq s \leq t}\big] = P_{u-t}(X_t,\,\cdot\,) \quad \text{a.s.} \qquad (0 \leq t \leq u). \tag{4.2}$$

Here we condition on the σ-field generated by the random variables $(X_s)_{0 \leq s \leq t}$. Formula (4.2) is equivalent to the statement that the finite-dimensional distributions of X are given by

$$\begin{aligned} &\mathbb{P}\big[X_0 \in dx_0, \ldots, X_{t_n} \in dx_n\big] \\ &\quad = \mathbb{P}[X_0 \in dx_0] P_{t_1 - t_0}(x_0, dx_1) \cdots P_{t_n - t_{n-1}}(x_{n-1}, dx_n) \end{aligned} \tag{4.3}$$

$(0 < t_1 < \cdots < t_n)$. Formula (4.3) is symbolic notation, which means that

$$\begin{aligned} &\mathbb{E}\big[f(X_0, \ldots, X_{t_n})\big] \\ &\quad = \int \mathbb{P}[X_0 \in dx_0] \int P_{t_1 - t_0}(x_0, dx_1) \cdots \int P_{t_n - t_{n-1}}(x_{n-1}, dx_n) f(x_0, \ldots, x_n) \end{aligned}$$

[6] The word cadlag is an abbreviation of the French **c**ontinue **à d**roit, **l**imite **à g**auche.

[7] It is possible to equip the space $\mathcal{D}_E[0,\infty)$ of cadlag functions $w \colon [0,\infty) \to E$ with a (rather natural) topology, called the *Skorohod topology*, such that $\mathcal{D}_E[0,\infty)$ is a Polish space and the Borel σ-field on $\mathcal{D}_E[0,\infty)$ is generated by the coordinate projections $w \mapsto w_t$ $(t \geq 0)$. As a result, we can view a stochastic process $X = (X_t)_{t \geq 0}$ with cadlag sample paths as a single random variable X taking values in the space $\mathcal{D}_E[0,\infty)$. The law of such a random variable is then uniquely determined by the finite-dimensional distributions of $(X_t)_{t \geq 0}$.

for all $f \in B(E^{n+1})$. By (4.3), the law of a Feller process X is uniquely determined by its initial law $\mathbb{P}[X_0 \in \cdot\,]$ and its transition probabilities $(P_t)_{t \geq 0}$. Existence is less obvious than uniqueness, but the next theorem says that this holds in full generality.

Theorem 4.2 (Construction of Feller processes) *Let E be a compact metrizable space, let μ be a probability measure on E, and let $(P_t)_{t\geq 0}$ be a Feller semigroup. Then there exists a Feller process $X = (X_t)_{t\geq 0}$ with initial law $\mathbb{P}[X_0 \in \cdot\,] = \mu$, and such a process is unique in distribution.*

Just as in the case for finite state space, we would like to characterize a Feller semigroup by its generator. This is somewhat more complicated than in the finite setting since, in general, it is not possible to make sense of the exponential formula $P_t = e^{tG} := \sum_{n=0}^{\infty} \frac{1}{n!}(tG)^n$. This is related to the fact that if G is the generator of a Feller semigroup, then in general it is not possible to define Gf for all $f \in C(E)$, as we now explain.

Let $\mathcal{V}$ be a Banach space. (In our case, the only Banach spaces that we will need are spaces of the form $C(E)$, equipped with the supremum norm.) By definition, a *linear operator* on $\mathcal{V}$ is a pair $(A, \mathcal{D}(A))$ where $\mathcal{D}(A)$ is a linear subspace of $\mathcal{V}$, called the *domain* and A is a linear map $A\colon \mathcal{D}(A) \to \mathcal{V}$. Even though a linear operator is really a pair $(A, \mathcal{D}(A))$, one often writes sentences such as "let A be a linear operator" without explicitly mentioning the domain. This is similar to phrases like: "let $\mathcal{V}$ be a Banach space" (without mentioning the norm) or "let M be a measurable space" (without mentioning the σ-field).

We say that a linear operator A (with domain $\mathcal{D}(A)$) on a Banach space $\mathcal{V}$ is *closed* if and only if its graph $\{(f, Af) : f \in \mathcal{D}(A)\}$ is a closed subset of $\mathcal{V} \times \mathcal{V}$. By definition, a linear operator A (with domain $\mathcal{D}(A)$) on a Banach space $\mathcal{V}$ is *closable* if the closure of its graph (as a subset of $\mathcal{V} \times \mathcal{V}$) is the graph of a linear operator $\overline{A}$ with domain $\mathcal{D}(\overline{A})$. This operator is then called the *closure* of A. We mention the following theorem.

Theorem 4.3 (Closed graph theorem) *Let $\mathcal{V}$ be a Banach space, and let A be a linear operator that is everywhere defined, that is, $\mathcal{D}(A) = \mathcal{V}$. Then the following statements are equivalent.*

(i) *A is continuous as a map from $\mathcal{V}$ into itself.*
(ii) *A is* bounded, *that is, there exists a constant $C < \infty$ such that $\|Af\| \leq C\|f\|$ $(f \in \mathcal{V})$.*
(iii) *A is closed.*

Theorem 4.3 shows in particular that if A is an unbounded operator (i.e., there exists $0 \neq f_n \in \mathcal{D}(A)$ such that $\|Af_n\|/\|f_n\| \to \infty$) and A is closable, then its closure $\overline{A}$ will not be everywhere defined. Closed (but possibly unbounded) linear operators are in a sense "the next good thing" after bounded operators.

As before, let E be a compact metrizable space, and let $(P_t)_{t\geq 0}$ be a continuous transition probability (or equivalently Feller semigroup) on E. By definition, the *generator* of $(P_t)_{t\geq 0}$ is the linear operator

$$Gf := \lim_{t\to 0} t^{-1}(P_t f - f),$$

with domain

$$\mathcal{D}(G) := \big\{f \in C(E) : \text{the limit } \lim_{t\to 0} t^{-1}(P_t f - f) \text{ exists}\big\}.$$

Here, when we say that the limit exists, we mean (of course) with respect to the topology on $C(E)$, that is, w.r.t. the supremum norm. The following lemma says that generators are closed, densely defined operators.

Lemma 4.4 (Elementary properties of generators) *Let G be the generator of a Feller semigroup $(P_t)_{t\geq 0}$. Then G is closed and $\mathcal{D}(G)$ is a dense subspace of $C(E)$.*

Since G is closed, Theorem 4.3 tells us that G is everywhere defined (i.e., $\mathcal{D}(G) = C(E)$) if and only if G is bounded. For bounded generators, it is not hard to show that the exponential formula $e^{tG}f := \sum_{n=0}^{\infty} \frac{1}{n!}(tG)^n f$ $(f \in C(E))$ converges in the norm on $C(E)$ and that the Feller semigroup with generator G is given by $P_t = e^{tG}$. On the other hand, if G is unbounded, then it is in general not possible to make sense of the exponential formula.[8] In the context of interacting particle systems, it is not hard to show that a generator of the form (4.1) is bounded if $\sum_{m\in\mathcal{G}} r_m < \infty$. For the particle systems we will be interested in, this sum will usually be infinite and the generator will be unbounded.

Since we cannot use the exponential formula $P_t = e^{tG}$, we need another way to characterize $(P_t)_{t\geq 0}$ in terms of G. Similar to what we did in Section 2.3, we will use the backward equation instead. Let A be a linear operator on $C(E)$. By definition, we say that a function $[0,\infty) \ni t \mapsto u_t \in C(E)$ solves the *Cauchy equation*

$$\tfrac{\partial}{\partial t} u_t = A u_t \qquad (t \geq 0) \tag{4.4}$$

if $u_t \in \mathcal{D}(A)$ for all $t \geq 0$, the maps $t \mapsto u_t$ and $t \mapsto Au_t$ are continuous (w.r.t. the topology on $C(E)$), the limit $\frac{\partial}{\partial t} u_t := \lim_{s\to 0} s^{-1}(u_{t+s} - u_s)$ exists (w.r.t. the topology on $C(E)$) for all $t \geq 0$, and (4.4) holds. The following proposition shows that a Feller semigroup is uniquely characterized by its generator.

Proposition 4.5 (Cauchy problem) *Let G be the generator of a Feller semigroup $(P_t)_{t\geq 0}$. Then, for each $f \in \mathcal{D}(G)$, the Cauchy equation $\frac{\partial}{\partial t} u_t = G u_t$ $(t \geq 0)$ has a unique solution $(u_t)_{t\geq 0}$ with initial state $u_0 = f$. Denoting this solution by $U_t f := u_t$ defines for each $t \geq 0$ a linear operator U_t with domain $\mathcal{D}(G)$, of which $P_t = \overline{U}_t$ is the closure.*

We need a way to check that (the closure of) a given operator is the generator of a Feller semigroup. For a given linear operator A, constant $\lambda > 0$, and $f \in C(E)$, we say that a function $p \in C(E)$ solves the *Laplace equation*

$$(\lambda - A)p = f \tag{4.5}$$

if $p \in \mathcal{D}(A)$ and (4.5) holds. The following lemma shows how solutions to Laplace equations typically arise.

Lemma 4.6 (Laplace equation) *Let G be the generator of a Feller semigroup $(P_t)_{t\geq 0}$ on $C(E)$, let $\lambda > 0$ and $f \in C(E)$. Then the Laplace equation $(\lambda - G)p = f$ has a unique solution, that is given by*

$$p = \int_0^\infty P_t f\, e^{-\lambda t}\,\mathrm{d}t.$$

[8] In order for $\sum_{n=0}^{\infty} \frac{1}{n!} t^n G^n f$ to make sense, we need that $G^n f$ is well-defined for all $n \geq 0$. For $n = 1$, this already requires that $f \in \mathcal{D}(G)$ but for higher n we need even more since it is in general not true that G maps $\mathcal{D}(G)$ into itself. Thus, it is not even clear for which class of functions we can make sense of each term in the expansion separately.

We say that an operator A on $C(E)$ with domain $\mathcal{D}(A)$ satisfies the *positive maximum principle* if, whenever a function $f \in \mathcal{D}(A)$ assumes its maximum over E in a point $x \in E$ and $f(x) \geq 0$, we have $Af(x) \leq 0$. The following proposition gives necessary and sufficient conditions for a linear operator G to be the generator of a Feller semigroup.

Theorem 4.7 (Generators of Feller semigroups) *A linear operator G on $C(E)$ is the generator of a Feller semigroup $(P_t)_{t\geq 0}$ if and only if*

(i) *$1 \in \mathcal{D}(G)$ and $G1 = 0$.*
(ii) *G satisfies the positive maximum principle.*
(iii) *$\mathcal{D}(G)$ is dense in $C(E)$.*
(iv) *For every $f \in C(E)$ and $\lambda > 0$, the Laplace equation $(\lambda - G)p = f$ has a solution.*

In practice, it is rarely possible to give an explicit description of the (full) domain of a Feller generator. Rather, one often starts with an operator that is defined on a smaller domain of "nice" functions and then takes its closure. In general, if G is a closed linear operator and $\mathcal{D}' \subset \mathcal{D}(G)$ is a linear subspace of $\mathcal{D}(G)$, then we let $G|_{\mathcal{D}'}$ denote the restriction of G to $\mathcal{D}'$, that is, $G|_{\mathcal{D}'}$ is the linear operator with domain $\mathcal{D}(G|_{\mathcal{D}'}) := \mathcal{D}'$ defined as $G|_{\mathcal{D}'}f := Gf$ for all $f \in \mathcal{D}'$. We say that $\mathcal{D}'$ is a *core* for G if $\overline{G|_{\mathcal{D}'}} = G$.

Lemma 4.8 (Core of a generator) *Let G be the generator of a Feller semigroup, and let $\mathcal{D}'$ be a linear subspace of $\mathcal{D}(G)$. Assume that $\mathcal{D}'$ is dense in $C(E)$. Then the following conditions are equivalent:*

(i) *$\mathcal{D}'$ is a core for G,*
(ii) *the set $\{(\lambda - A)p : p \in \mathcal{D}'\}$ is dense in $C(E)$ for some $\lambda > 0$,*
(iii) *the set $\{(\lambda - A)p : p \in \mathcal{D}'\}$ is dense in $C(E)$ for all $\lambda > 0$.*

Note that by condition (ii) of Lemma 4.8, to check that a dense set $\mathcal{D}' \subset C(E)$ is a core for G, it suffices to show that for some $\lambda > 0$, there exists a dense subspace $\mathcal{R} \subset C(E)$ such that for every $f \in \mathcal{R}$, the Laplace equation $(\lambda - A)p = f$ has a solution $p \in \mathcal{D}'$. Using Lemma 4.8, one can prove the following version of the Hille–Yosida theorem.

Theorem 4.9 (Hille–Yosida) *A linear operator A on $C(E)$ with domain $\mathcal{D}(A)$ is closable and its closure $G := \overline{A}$ is the generator of a Feller semigroup if and only if*

(i) *There exist $f_n \in \mathcal{D}(A)$ such that $f_n \to 1$ and $Af_n \to 0$.*
(ii) *A satisfies the positive maximum principle.*
(iii) *$\mathcal{D}(A)$ is dense in $C(E)$.*
(iv) *For some (and hence for all) $\lambda \in (0, \infty)$, there exists a dense subspace $\mathcal{R} \subset C(E)$ such that for every $f \in \mathcal{R}$, the Laplace equation $(\lambda - A)p = f$ has a solution p.*

Conditions (i)–(iii) are usually easy to verify for a given operator A, but condition (iv) is the "hard" condition since this means that one has to prove existence of solutions to the Laplace equation $(\lambda - G)p = f$ for a dense set of functions f.

If K is a probability kernel on E and $r > 0$, then

$$Gf := r(Kf - f) \qquad \big(f \in C(E)\big) \tag{4.6}$$

defines a Feller generator that is everywhere defined (i.e., $\mathcal{D}(G) = C(E)$) and hence, in view of Theorem 4.3, a bounded operator. For generators of this simple form, one can construct the corresponding semigroup by the exponential formula

$$P_t f = e^{tG} f := \sum_{n=0}^{\infty} \frac{1}{n!}(tG)^n f,$$

where the infinite sum converges in $C(E)$. The corresponding Markov process has a simple description: with rate r, the process jumps from its current position x to a new position chosen according to the probability law $K(x\,\cdot)$.

As soon as Feller processes get more complicated in the sense that "the total rate of all things that can happen" is infinite (as will be the case for interacting particle systems), one needs the more complicated Hille–Yosida theory. To demonstrate the strength of Theorem 4.9, consider $E := [0,1]$ and the linear operator A defined by $\mathcal{D}(A) := C^2[0,1]$ (the space of twice continuously differentiable functions on $[0,1]$) and

$$Af(x) := x(1-x)\tfrac{\partial^2}{\partial x^2} f(x) \qquad \big(x \in [0,1]\big). \tag{4.7}$$

One can show that A satisfies the conditions of Theorem 4.9, and hence $\overline{A}$ generates a Feller semigroup. The corresponding Markov process turns out to have continuous sample paths and is indeed the *Wright–Fisher diffusion* that we met before in formula (3.26).

Exercise 4.10 (Brownian motion) *Let $(P_t)_{t\geq 0}$ denote the transition kernels of Brownian motion on $\mathbb{R}^d$. Let $E := \mathbb{R}^d \cup \{\infty\}$ denote the one-point compactification of $\mathbb{R}^d$ and extend P_t $(t \geq 0)$ to probability kernels on E by setting $P_t(\infty,\,\cdot\,) := \delta_\infty$. Show that $(P_t)_{t\geq 0}$ is a Feller semigroup.*

In Chapter 2, we viewed (possibly explosive) continuous-time Markov chains with a countable state space S as Markov processes on the extended state space S_∞, where S_∞ is the one-point compactification of S. It is natural to ask if they are in fact Feller processes on S_∞. The answer is, in general, negative. The reason is that the extended transition kernels $(\overline{P}_t)_{t\geq 0}$ on S_∞ may fail to be continuous at ∞, that is, $\overline{P}_t(x_n,\,\cdot\,)$ may fail to converge to $\overline{P}_t(\infty,\,\cdot\,)$ if $x_n \to \infty$. In many cases where this problem occurs, it can be solved by choosing another compactification of S (i.e., by adding more points at infinity). Whether this can be done in general, I don't know.

Exercise 4.11 (Wright–Fisher diffusion) *Show that the operator A defined in (4.7) satisfies the conditions of Theorem 4.9.* Hint: *Show that if f is a polynomial of order n, then so is Af. Use this to show that the Cauchy equation $\frac{\partial}{\partial t}u_t = Au_t$ has a solution for each initial state $u_0 = f$ that is a polynomial. Then show that $p := \int_0^\infty u_t\, e^{-\lambda t}\,\mathrm{d}t$ solves the Laplace equation $(\lambda - A)p = f$.*

Some Notes on the Proofs

In the remainder of this section, we indicate where proofs of the stated theorems can be found. Readers who are more interested in interacting particle systems than in functional analysis may skip from here to the next section.

The fact that there is a one-to-one correspondence between continuous transition probabilities and collections $(P_t)_{t\geq 0}$ of linear operators satisfying the assumptions (i)–(iv) of a Feller semigroup follows from [Kal97, Prop. 17.14].

Theorem 4.2 (including a proof) can be found in [Kal97, Thm 17.15] and [EK86, Thm 4.2.7]. Theorem 4.3 (the closed graph theorem and characterization of continuous linear maps) can be found in many places (including Wikipedia).

Lemma 4.4 follows from [EK86, Corollary I.1.6]. The statements of this lemma can also easily be derived from the Hille–Yosida theorem (see below). Proposition 4.5 summarizes a number of well-known facts. The fact that $u_t := P_t f$ solves the Cauchy equation if $f \in \mathcal{D}(G)$ is proved in [EK86, Prop 1.1.5 (b)], [Kal97, Thm 17.6], and [Lig10, Thm 3.16 (b)]. To see that solutions to the Cauchy equation are unique, we use the following fact.

Lemma 4.12 (Positive maximum principle) *Let A be a linear operator on $C(E)$ and let $u = (u_t)_{t\geq 0}$ be a solution to the Cauchy equation $\frac{\partial}{\partial t}u_t = Au_t$ $(t \geq 0)$. Assume that A satisfies the positive maximum principle and $u_0 \geq 0$. Then $u_t \geq 0$ for all $t \geq 0$.*

Proof By linearity, we may equivalently show that $u_0 \leq 0$ implies $u_t \leq 0$. Assume that $u_t(x) > 0$ for some $x \in E$. By the compactness of E, the function $(x,t) \mapsto e^{-t}u_t(x)$ must assume its maximum over $E \times [0,t]$ in some point (y,s). Since u is positive somewhere on $E \times [0,t]$ we have $e^{-s}u_s(y) > 0$, and hence $s > 0$ by the fact that $u_0 \leq 0$. But now, since A satisfies the positive maximum principle,

$$0 \leq \frac{\partial}{\partial s}\big(e^{-s}u_s(y)\big) = -e^{-s}u_s(y) + e^{-s}Au_s(y) \leq -e^{-s}u_s(y) < 0,$$

so we arrive at a contradiction. ∎

By linearity, Lemma 4.12 implies that if u, v are two solutions to the same Cauchy equation and $u_0 \leq v_0$, then $u_t \leq v_t$ for all $t \geq 0$. In particular, since by Theorem 4.7, Feller generators satisfy the positive maximum principle, this implies uniqueness of solutions of the Cauchy equation in Proposition 4.5. Again by Theorem 4.7, the domain of a Feller semigroup is a dense subspace of $C(E)$, so the final statement of Proposition 4.5 follows from the following simple lemma and the fact that $\|P_t f\|_\infty \leq \|f\|_\infty$.

Lemma 4.13 (Closure of bounded operators) *Let $(\mathcal{V}, \|\cdot\|)$ be a Banach space, and let A be a linear operator on $\mathcal{V}$ such that $\mathcal{D}(A)$ is dense and $\|Af\| \leq C\|f\|$ $(f \in \mathcal{D}(A))$ for some $C < \infty$. Then A is closable, $\mathcal{D}(\overline{A}) = \mathcal{V}$, and $\|\overline{A}f\| \leq C\|f\|$ $(f \in \mathcal{V})$.*

Proof (sketch) Since $\mathcal{D}(A)$ is dense, for each $f \in \mathcal{V}$ we can choose $\mathcal{D}(A) \ni f_n \to f$. Using the fact that A is bounded, it is easy to check that if $(f_n)_{n\geq 0}$ is a Cauchy sequence and $f_n \in \mathcal{D}(A)$ for all n, then $(Af_n)_{n\geq 0}$ is also a Cauchy sequence. By the completeness of $\mathcal{V}$, it follows that the limit $\overline{A}f := \lim_{n\to\infty} Af_n$ exists for all $f \in \mathcal{V}$. To see that this defines $\overline{A}$ unambiguously, assume that $f_n \to f$ and $g_n \to f$ and observe that $\|Af_n - Ag_n\| \leq C\|f_n - g_n\| \to 0$. The fact that $\|\overline{A}f\| \leq C\|f\|$ $(f \in \mathcal{V})$ follows from the continuity of the norm. ∎

Lemma 4.6 follows from [EK86, Prop 1.2.1]. Theorems 4.7 and 4.9 both go under the name of the Hille–Yosida theorem. Often, they are stated in a more general form without condition (i). In this generality, the operator G generates a semigroup of *subprobability kernels* $(P_t)_{t\geq 0}$, that is, $P_t(x, \cdot)$ is a measure with total mass $P_t(x,E) \leq 1$. In this context, a Feller semigroup with $P_t(x, E) = 1$ for all t, x is called *conservative*. It is clear from Proposition 4.5 that condition (i) in Theorems 4.7 and 4.9 is necessary and sufficient for the Feller group to be conservative.

The versions of the Hille–Yosida theorem stated in [EK86, Kal97] are more general than Theorems 4.7 and 4.9 since they allow for the case that E is not compact but only locally compact. This is not really more general, however, since what these books basically do if E is not compact is the following. First, they construct the one-point compactification $\overline{E} = E \cup \{\infty\}$ of E. Next, they extend the transition probabilities to $\overline{E}$ by putting $P_t(\infty, \cdot\,) := \delta_\infty$ for all $t \geq 0$. Having proved that they generate a conservative Feller semigroup on $\overline{E}$ of this form, they then still need to prove that the associated Markov process does not explode in the sense that $\mathbb{P}^x[X_t \in E \ \forall t \geq 0] = 1$ for all $x \in E$. In practical situations (such as when constructing Markov processes with state space $\mathbb{R}^d$) it is usually better to explicitly work with the one-point compactification of $\mathbb{R}^d$ instead of trying to formulate theorems for locally compact spaces that try to hide this compactification in the background.

Theorems 4.7 and 4.9 are special cases of more general theorems (also called Hille–Yosida theorem) for strongly continuous contraction semigroups taking values in a general Banach space. In this context, the positive maximum principle is replaced by the assumption that the operator under consideration is *dissipative*. In this more general setting, Theorems 4.7 and 4.9 correspond to [EK86, Thms 1.2.6 and 1.2.12]. Lemma 4.8 follows from [EK86, Lemma 1.2.11 and Prop 1.3.1]. In the more specific setup of Feller semigroups, versions of Theorem 4.9 can be found in [EK86, Thm 4.2.2] and [Kal97, Thm 17.11]. There is also an account of Hille–Yosida theory for Feller semigroups in [Lig10, Chap 3], but this reference does not mention the positive maximum principle (using a dissipativity assumption instead).

Feller semigroups with bounded generators such as in (4.6) are treated in [EK86, Sect 4.2] and [Kal97, Prop 17.2].

4.3 Poisson Construction

We briefly recall the setup introduced in Section 4.1. S is a finite set, called the *local state space*, and Λ is a countable set, called the *lattice*. We equip the product space S^Λ with the product topology, making it into a compact metrizable space. Elements of S^Λ are denoted $x = (x(i))_{i \in \Lambda}$. We fix a countable set $\mathcal{G}$ whose elements are continuous maps $m \colon S^\Lambda \to S^\Lambda$ as well as nonnegative rates $(r_m)_{m \in \mathcal{G}}$. Our aim is to construct a Markov process with formal generator of the form (4.1), that is,

$$Gf(x) = \sum_{m \in \mathcal{G}} r_m \{ f(m(x)) - f(x) \} \qquad (x \in S^\Lambda).$$

Our approach is very similar to the Poisson construction of continuous-time Markov chains described in Section 2.6. We equip the space $\mathcal{G} \times \mathbb{R}$ with the measure

$$\rho(\{m\} \times [s, t]) := r_m (t - s) \qquad (m \in \mathcal{G},\ s \leq t). \tag{4.8}$$

Let ω be a Poisson point set with intensity ρ. We call ω the *graphical representation* associated with the random mapping representation (2.21). Since $\mathcal{G}$ is countable, by the argument used in Section 2.6, the time coordinates of points $(m, t) \in \omega$ are all different.

Therefore, as we did in the case of continuous-time Markov chains, we can unambiguously define a random function $\mathbb{R} \ni t \mapsto \mathfrak{m}_t^\omega \in \mathcal{G}$ by setting

$$\mathfrak{m}_t^\omega := \begin{cases} m & \text{if } (m,t) \in \omega, \\ 1 & \text{otherwise,} \end{cases} \tag{4.9}$$

where we write 1 to denote the identity map. By definition, we say that a random function $X\colon [s,\infty) \to S^\Lambda$ solves the evolution equation

$$X_t = \mathfrak{m}_t^\omega(X_{t-}) \qquad (t > s), \tag{4.10}$$

if $[s,\infty) \ni t \mapsto X_t \in S^\Lambda$ is cadlag and (4.10) holds. We recall that for any continuous map m and site $i \in \Lambda$, the set $\mathcal{R}_i^\downarrow(m)$ has been defined in Section 2.9. Here is the main result of the present section.

Theorem 4.14 (Poisson construction) *Assume that the rates* $(r_m)_{m\in\mathcal{G}}$ *satisfy*

$$\text{(i)}\ \sup_{i\in\Lambda} \sum_{m\in\mathcal{G}} r_m 1_{\mathcal{D}(m)}(i) < \infty, \quad \text{(ii)}\ \sup_{i\in\Lambda} \sum_{m\in\mathcal{G}} r_m \big|\mathcal{R}_i^\downarrow(m)\backslash\{i\}\big| < \infty. \tag{4.11}$$

Then almost surely, for each $s \in \mathbb{R}$ *and* $x \in S^\Lambda$, *there exists a unique solution* $(X_t^{s,x})_{t\geq s}$ *to the evolution equation (4.10) with initial state* $X_s^{s,x} = x$. *Setting*

$$\mathbf{X}_{s,t}(x) := X_t^{s,x} \qquad (s \leq t,\ x \in S^\Lambda) \tag{4.12}$$

defines a collection of continuous maps $(\mathbf{X}_{s,t})_{s\leq t}$ *from* S^Λ *into itself such that*

$$\mathbf{X}_{s,s} = 1 \quad \text{and} \quad \mathbf{X}_{t,u} \circ \mathbf{X}_{s,t} = \mathbf{X}_{s,u} \qquad (s \leq t \leq u). \tag{4.13}$$

Setting

$$P_t(x,\,\cdot\,) := \mathbb{P}\big[\mathbf{X}_{0,t}(x) \in \cdot\,\big] \qquad (t \geq 0,\ x \in S^\Lambda) \tag{4.14}$$

defines the semigroup of a Feller process with state space S^Λ. *If* $s \in \mathbb{R}$ *and* X_0 *is an* S^Λ*-valued random variable with law* μ, *independent of* ω, *then the process* $(X_t)_{t\geq 0}$ *defined as*

$$X_t := \mathbf{X}_{s,s+t}(X_0) \qquad (t \geq 0) \tag{4.15}$$

is distributed as the Feller process with semigroup $(P_t)_{t\geq 0}$ *and initial law* μ.

One may notice the similarity between condition (4.11) and the condition (2.41) from Chapter 2, the only difference being that $\mathcal{R}_i^\uparrow(m)$ in condition (2.41) (ii) is replaced by $\mathcal{R}_i^\downarrow(m)$, which looks "downwards" in time. (Here we use our usual convention of plotting time upwards in pictures of graphical representations so that downwards means back in time.) In Proposition 4.24 of Section 4.4, we will see that the "upward" condition (2.41) (ii) in general guarantees that finite perturbations of the initial state have finite consequences at later times. For later use, we introduce three constants whose finiteness is guaranteed by conditions (4.11) (ii) and (2.41) (ii) and by part (i) of either of these equations.

$$\begin{gathered} K_\downarrow := \sup_{i\in\Lambda} \sum_{m\in\mathcal{G}} r_m\big(|\mathcal{R}_i^\downarrow(m)| - 1\big), \qquad K_\uparrow := \sup_{i\in\Lambda} \sum_{m\in\mathcal{G}} r_m\big(|\mathcal{R}_i^\uparrow(m)| - 1\big), \\ K_0 := \sup_{i\in\Lambda} \sum_{m\in\mathcal{G}} r_m 1_{\mathcal{D}(m)}(i). \end{gathered} \tag{4.16}$$

The proof of Theorem 4.14 will take up the rest of this section. At first, it may be surprising that solutions of the evolution equation (4.10) with a given initial state are unique. After all, if we replace the compact set S^Λ by $[0, 1]$, then there are many cadlag functions $X\colon [s, \infty) \to [0, 1]$ with a given initial state that make no jumps at all. The following exercise shows that at least in the case when $\omega = \emptyset$, the equation (4.10) has a unique solution.

Exercise 4.15 (Total disconnectedness) *A topological space E is* totally disconnected *if for each $x_1, x_2 \in E$ with $x_1 \neq x_2$, there exist open sets $O_1 \ni x_1$ and $O_2 \ni x_2$ such that $O_1 \cap O_2 = \emptyset$ and $O_1 \cup O_2 = E$. Prove that S^Λ is totally disconnected. Prove that if E is a totally disconnected space, then each continuous function $f\colon [0, \infty) \to E$ is constant.*

The difficulty with proving that for each $s \in \mathbb{R}$ and $x \in S^\Lambda$, the equation (4.10) has a unique solution $X^{s,x}$ is that typically condition (2.20) will be violated. As a result, $\{t > s : (m, t) \in \omega,\ m(x) \neq x\}$ is a dense subset of $[s, \infty)$ and solutions to (4.10) will not be piecewise constant. However, most of the jumps of $(X_t^{s,x})_{t \geq s}$ will involve sites that are far away, and $t \mapsto X_t^{s,x}(i)$ will still be piecewise constant for each fixed $i \in \Lambda$.

The trick for proving uniqueness of solutions of (4.10) is to look backwards in time. We recall from Lemma 2.30 that if T is a finite set and $f\colon S^\Lambda \to T$ is a continuous function, then f depends on finitely many coordinates. A consequence of this is that the space $C(S^\Lambda, T)$ of continuous functions $\phi\colon S^\Lambda \to T$ is countable. It turns out that if we fix $\phi \in C(S^\Lambda, T)$ and a time $u \in \mathbb{R}$, then the stochastic flow $(\mathbf{X}_{s,t})_{s \leq t}$ that we are about to construct has the property that the process

$$\Phi_t := \phi \circ \mathbf{X}_{u-t,u} \qquad (t \geq 0)$$

is a nonexplosive continuous-time Markov chain with countable state space $C(S^\Lambda, T)$. We call $(\Phi_t)_{t \geq 0}$ the *backward in time process*. The first step toward proving Theorem 4.14 is showing that this continuous-time Markov chain is nonexplosive.

Proposition 4.16 (Backward in time process) *Assume (4.11). Let T be a finite set. Then setting*

$$Hf(\phi) := \sum_{m \in \mathcal{G}} r_m \big\{ f(\phi \circ m) - f(\phi) \big\} \tag{4.17}$$

for all bounded $f\colon C(S^\Lambda, T) \to [0, \infty)$ defines the generator of a nonexplosive continuous-time Markov chain $(\Phi_t)_{t \geq 0}$ with state space $C(S^\Lambda, T)$. This Markov chain satisfies

$$\mathbb{E}^\phi\big[|\mathcal{R}(\Phi_t)|\big] \leq |\mathcal{R}(\phi)| e^{K_\downarrow t} \qquad \big(t \geq 0,\ \phi \in C(S^\Lambda, T)\big), \tag{4.18}$$

where $K_\downarrow$ is defined in (4.16).

Before we prove Proposition 4.16, we first discuss its consequences. Combining Proposition 4.16 with Theorem 2.24, one obtains that almost surely, for each $u \in \mathbb{R}$ and $\phi \in C(S^\Lambda, T)$, there exists a unique cadlag function $\Phi^{u,\phi}\colon (-\infty, u] \to C(S^\Lambda, T)$ such that $\Phi_u^{u,\phi} = \phi$ and

$$\Phi_{t-}^{u,\phi} = \begin{cases} \Phi_t^{u,\phi} \circ m & \text{if } (m, t) \in \omega, \\ \Phi_t^{u,\phi} & \text{otherwise} \end{cases} \qquad (t \leq u). \tag{4.19}$$

Setting

$$\mathbf{F}_{u,t}(\phi) := \Phi_t^{u,\phi} \qquad \big(t \leq u,\ \phi \in C(S^\Lambda, T)\big) \tag{4.20}$$

defines a collection of maps $(\mathbf{F}_{u,t})_{u \geq t}$ from $C(S^\Lambda, T)$ into itself such that

$$\mathbf{F}_{u,u} = 1 \quad \text{and} \quad \mathbf{F}_{t,s} \circ \mathbf{F}_{u,t} = \mathbf{F}_{u,s} \qquad (u \geq t \geq s).$$

If $u \in \mathbb{R}$ and Φ_0 is a $C(S^\Lambda, T)$-valued random variable with law μ, independent of ω, then the process $(\Phi_t)_{t \geq 0}$ defined as

$$\Phi_t := \mathbf{F}_{u,u-t}(\Phi_0) \qquad (t \geq 0)$$

is distributed as the left-continuous modification of the continuous-time Markov chain with generator H from (4.17) and initial law μ. These facts are illustrated in Figure 4.1.

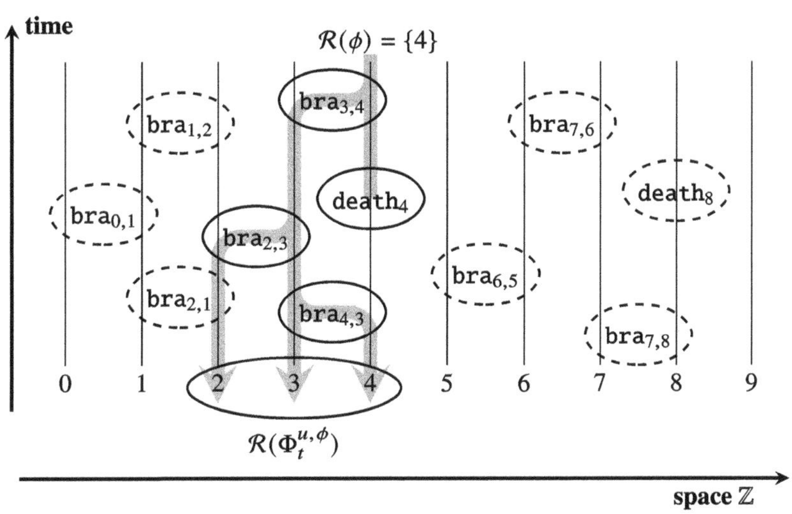

Figure 4.1 Graphical representation of a one-dimensional contact process, with the backwards in time process $(\Phi_t^{u,\phi})_{t \in (-\infty,u]}$. In this example, $T = S = \{0, 1\}$, $\phi \in C(S^\Lambda, T)$ is defined by $\phi(x) := x(4)$ $(x \in S^\Lambda)$, and one has $\Phi_t^{u,\phi} = \phi \circ \mathtt{bra}_{3,4} \circ \mathtt{death}_4 \circ \mathtt{bra}_{2,3} \circ \mathtt{bra}_{4,3}$. The gray arrows pointing downwards show the evolution backwards in time of the set $\mathcal{R}(\Phi_t^{u,\phi})$ of relevant lattice points for the map $\Phi_t^{u,\phi}$.

The proof of Proposition 4.16 needs one preparatory lemma.

Lemma 4.17 (Well-defined generator) *Assume that the rates $(r_m)_{m \in \mathcal{G}}$ satisfy*

$$\sum_{m \in \mathcal{G}} 1_{\mathcal{D}(m)}(i) r_m < \infty \quad \forall i \in \Lambda. \tag{4.21}$$

Then formula (4.17) defines the generator of a (possibly explosive) continuous-time Markov chain $(\Phi_t)_{t \geq 0}$ with state space $C(S^\Lambda, T)$.

Proof We must check condition (2.20), which in the present context says that

$$c(\phi) := \sum_{m:\, \phi \circ m \neq \phi} r_m < \infty \quad \text{for all } \phi \in C(S^\Lambda, T).$$

We observe that $\phi(m(x)) \neq \phi(x)$ for some $x \in S^\Lambda$ implies that $\mathcal{D}(m) \cap \mathcal{R}(\phi) \neq \emptyset$, so we can estimate

$$\sum_{m:\,\phi\circ m\neq\phi} r_m \le \sum_{m:\,\mathcal{D}(m)\cap\mathcal{R}(\phi)\neq\emptyset} r_m \le \sum_{i\in\mathcal{R}(\phi)} \sum_m 1_{\mathcal{D}(m)}(i) r_m, \tag{4.22}$$

which is finite by (4.21) and the finiteness of $\mathcal{R}(\phi)$. ∎

Proof of Proposition 4.16 Condition (4.11) (i) clearly implies (4.21), so by Lemma 4.17 H is the generator of a (possibly explosive) continuous-time Markov chain with state space $\mathcal{C}(S^\Lambda, T)$. To prove that H is nonexplosive, we apply Theorem 2.12 to the Lyapunov function

$$L(\phi) := |\mathcal{R}(\phi)| \qquad (\phi \in \mathcal{C}(S^\Lambda, T)).$$

Formula (4.22) shows that

$$\sup\{c(\phi) : \phi \in \mathcal{C}(S^\Lambda, T),\ L(\phi) < C\} \le C \sup_{i\in\Lambda} \sum_m 1_{\mathcal{D}(m)}(i) r_m,$$

which by (4.11) (i) implies that L satisfies condition (i) of Theorem 2.12. It remains to check condition (ii). We will show that $HL \le K_\downarrow L$, where $K_\downarrow$ is the constant defined in (4.16), which is finite by (4.11) (ii). We observe that

$$HL(\phi) = \sum_{m\in\mathcal{G}} r_m \{L(\phi\circ m) - L(\phi)\} = \sum_{m\in\mathcal{G}} r_m \{|\mathcal{R}(\phi\circ m)| - |\mathcal{R}(\phi)|\}.$$

Since

$$\mathcal{R}(\phi\circ m) \subset \bigcup_{i\in\mathcal{R}(\phi)} \mathcal{R}_i^\downarrow(m),$$

we can estimate

$$|\mathcal{R}(\phi\circ m)| - |\mathcal{R}(\phi)| \le \sum_{i\in\mathcal{R}(\phi)} \big(|\mathcal{R}_i^\downarrow(m)| - 1\big).$$

It follows that

$$HL(\phi) \le \sum_{m\in\mathcal{G}} \sum_{i\in\mathcal{R}(\phi)} r_m \big(|\mathcal{R}_i^\downarrow(m)| - 1\big) \le K_\downarrow \big|\mathcal{R}(\phi)\big| = K_\downarrow L(\phi),$$

so condition (ii) of Theorem 2.12 is satisfied with $\lambda = K_\downarrow$. It follows that H is nonexplosive and (4.18) holds. ∎

Proposition 4.16 is the cornerstone of the proof of Theorem 4.14. We will prove the following result, which is sometimes applicable even when condition (4.11) of Theorem 4.14 is not satisfied. In Exercise 6.10, we will see that if the continuous-time Markov chain with generator H from (4.17) is explosive, then solutions to the evolution equation (4.10) may fail to be unique.

Theorem 4.18 (Graphical construction) *Assume that the rates $(r_m)_{m\in\mathcal{G}}$ satisfy (4.21) and that the continuous-time Markov chain with generator H from (4.17) is nonexplosive. Then the conclusions of Theorem 4.14 remain true.*

It turns out that the condition (4.11) is more or less optimal in a translation-invariant setting but not necessarily for inhomogeneous systems as we now explain. For each bijection $\psi\colon \Lambda \to \Lambda$ and map $m\colon S^\Lambda \to S^\Lambda$, we define a translated map $T_\psi m\colon S^\Lambda \to S^\Lambda$ by

$$T_\psi m(x) := m\big(x \circ \psi^{-1}\big) \qquad (x \in S^\Lambda),$$

where $x \circ \psi^{-1}$ denotes the concatenation of the functions $\psi^{-1}\colon \Lambda \to \Lambda$ and $x\colon \Lambda \to S$. Let $\mathcal{G}$ be a collection of continuous maps, and let $(r_m)_{m\in\mathcal{G}}$ be nonnegative rates. By definition, an *automorphism* of $(r_m)_{m\in\mathcal{G}}$ is a bijection $\psi\colon \Lambda \to \Lambda$ such that $T_\psi m \in \mathcal{G}$ for all $m \in \mathcal{G}$ and

$$r_{T_\psi m} = r_m \qquad (m \in \mathcal{G}).$$

We say that the rates $(r_m)_{m\in\mathcal{G}}$ are *transitive* if for each $i, j \in \Lambda$, there exists an automorphism ψ of $(r_m)_{m\in\mathcal{G}}$ such that $\psi(i) = \psi(j)$. In such a situation, we also say that the associated random mapping representation of a generator as in (4.1) is *transitive*. For transitive random mapping representations, the expressions

$$\text{(i)} \sum_{m\in\mathcal{G}} r_m 1_{\mathcal{D}(m)}(i) \quad \text{and} \quad \text{(ii)} \sum_{m\in\mathcal{G}} r_m \big|\mathcal{R}_i^{\downarrow}(m)\backslash\{i\}\big|$$

do not depend on $i \in \Lambda$ and the suprema in (4.11) (i) and (ii) can be dropped. In such situations, one can show that the condition (4.11) is more or less optimal. If Λ is a transitive graph, then the random mapping representations of the voter model, the contact process, and other systems on Λ that we have already seen are transitive. On the other hand, it is nowadays common to study interacting particle systems in a random environment, for example, on random graphs. The random mapping representations of such systems are clearly not transitive. In such situations, the supremum in (4.11) is a nuisance, and Theorem 4.18 may be applicable even when Theorem 4.14 is not applicable.

We now set out to prove Theorem 4.18, which by Proposition 4.16 implies Theorem 4.14.

Lemma 4.19 (Evolution equation) *Under the assumptions of Theorem 4.18, almost surely, for each $s \in \mathbb{R}$ and $x \in S^\Lambda$, the evolution equation (4.10) has a unique solution $(X_t^{s,x})_{t\ge s}$ with initial state $X_s^{s,x} = x$. For any finite set T and $\phi \in \mathcal{C}(S^\Lambda, T)$, this solution satisfies*

$$\phi(X_t^{s,x}) = \mathbf{F}_{t,s}(\phi)(x) \qquad (t \ge 0), \tag{4.23}$$

where $(\mathbf{F}_{t,s})_{t\ge s}$ is the backward stochastic flow defined in (4.20).

Proof By Theorem 2.24, the assumptions of Theorem 4.18 guarantee that the backward stochastic flow in (4.20) is well-defined. For each $i \in \Lambda$, we define $\phi_i \in \mathcal{C}(S^\Lambda, S)$ by $\phi_i(x) := x(i)$ $(x \in S^\Lambda)$. We now fix $s \in \mathbb{R}$ and $x \in S^\Lambda$ and define $(X_t)_{t\ge s}$ by

$$X_t(i) := \mathbf{F}_{t,s}(\phi_i)(x) \qquad (t \ge s,\ i \in \Lambda),$$

where $(\mathbf{F}_{t,s})_{t\ge s}$ is the backward stochastic flow defined in (4.20). Then $X_s = x$. Moreover, $t \mapsto X_t(i)$ is piecewise constant, right-continuous, and

$$X_t(i) = \begin{cases} m(X_{t-})(i) & \text{if } \exists (m,t) \in \omega \text{ s.t. } \phi_i \circ m \neq \phi_i, \\ X_{t-}(i) & \text{otherwise.} \end{cases}$$

Using (4.21) and the fact that $\phi_i \circ m \neq \phi_i$ implies $i \in \mathcal{D}(m)$, we see that $(X_t)_{t\ge s}$ is cadlag and solves (4.10). This establishes existence of solutions. To prove uniqueness, assume that

$(X_t)_{t\geq s}$ is cadlag and solves (4.10) with $X_s = x$. Let T be a finite set and fix $\phi \in C(S^\Lambda, T)$. We claim that for each $u > s$, the function

$$[s,u] \ni t \mapsto \mathbf{F}_{u,t}(\phi)(X_t) \tag{4.24}$$

is constant. Indeed, $t \mapsto \mathbf{F}_{u,t}(\phi)$ is piecewise constant and right-continuous and takes values in $C(S^\Lambda, T)$, so by the fact that functions in $C(S^\Lambda, T)$ depend on finitely many coordinates and $t \mapsto X_t(i)$ is piecewise constant and right-continuous for each $i \in \Lambda$, we see that also $t \mapsto \mathbf{F}_{u,t}(\phi)(X_t)$ is piecewise constant and right-continuous. For each $(m,t) \in \omega$ with $t \in (s,u]$, one has

$$\mathbf{F}_{u,t-}(\phi)(X_{t-}) = \mathbf{F}_{u,t}(\phi) \circ m(X_{t-}) = \mathbf{F}_{u,t}(\phi)(X_t),$$

while at the remaining times in $(s,u]$ trivially $\mathbf{F}_{u,t-}(\phi)(X_{t-}) = \mathbf{F}_{u,t}(\phi)(X_t)$. This proves that $t \mapsto \mathbf{F}_{u,t}(\phi)(X_t)$ is constant on $(s,u]$, and hence, by right-continuity, also on $[s,u]$. Since the function in (4.24) is constant,

$$\phi(X_u) = \mathbf{F}_{u,u}(\phi)(X_u) = \mathbf{F}_{u,s}(\phi)(X_s) = \mathbf{F}_{u,s}(\phi)(x).$$

Since this holds for arbitrary T and $\phi \in C(S^\Lambda, T)$, we conclude that $(X_t)_{t\geq s}$ is unique and that (4.23) holds. ∎

By Lemma 4.19, under the assumptions of Theorem 4.18, almost surely, for each $s \in \mathbb{R}$ and $x \in S^\Lambda$, the evolution equation (4.10) has a unique solution $(X_t^{s,x})_{t\geq s}$ with initial state $X_s^{s,x} = x$. We use this to define random maps $(\mathbf{X}_{s,t})_{s\leq t}$ as in (4.12). Then (4.23) implies that for each finite set T, one has

$$\mathbf{F}_{t,s}(\phi) = \phi \circ \mathbf{X}_{s,t} \qquad \big(s \leq t,\ \phi \in C(S^\Lambda, T)\big). \tag{4.25}$$

It is straightforward from the definition that these maps satisfy (4.13). The stochastic flow $(\mathbf{X}_{s,t})_{s\leq t}$ is clearly stationary. Using the fact that restrictions of a Poisson point set to disjoint parts of the space are independent, we also see that $(\mathbf{X}_{s,t})_{s\leq t}$ has independent increments.

Lemma 4.20 (Continuity of the flow) *Under the assumptions of Theorem 4.18, almost surely, the maps $\mathbf{X}_{s,t} \colon S^\Lambda \to S^\Lambda$ are continuous for all $s \leq t$.*

Proof Since S^Λ is equipped with the product topology, it suffices to show that $x \mapsto \mathbf{X}_{s,t}(x)(i)$ is continuous for all $s \leq t$ and $i \in \Lambda$. Using notation as in the proof of Lemma 4.19, we have by (4.25) that

$$\mathbf{X}_{s,t}(x)(i) = \phi_i\big(\mathbf{X}_{s,t}(x)\big) = \mathbf{F}_{t,s}(\phi_i)(x). \tag{4.26}$$

Since $\mathbf{F}_{t,s}(\phi_i) \in C(S^\Lambda, S)$, the map $x \mapsto \mathbf{F}_{t,s}(\phi_i)(x)$ is continuous. ∎

Lemma 4.21 (Almost sure continuity) *Assume that $x_n, x \in S^\Lambda$ satisfy $x_n \to x$ in the product topology and that $t_n, t \geq 0$ satisfy $t_n \to t$. Then, under the assumptions of Theorem 4.18,*

$$\mathbf{X}_{0,t_n}(x_n) \underset{n\to\infty}{\longrightarrow} \mathbf{X}_{0,t}(x) \quad \text{a.s.}$$

Proof In line with the notation introduced in Section 2.8, let $\mathbf{X}_{s,u}[i] \colon S^\Lambda \to S$ be defined as $\mathbf{X}_{s,u}[i](x) := \mathbf{X}_{s,u}(x)(i)$. Since S^Λ is equipped with the product topology, it suffices to show that $\mathbf{X}_{0,t_n}(x_n)(i)$ converges to $\mathbf{X}_{0,t}(x)(i)$ for each $i \in \Lambda$. Since t is deterministic, by (4.21) there a.s. exists an $\varepsilon > 0$ such that

$$i \notin \mathcal{D}(m) \quad \forall (m,r) \in \omega \quad \text{with} \quad t - \varepsilon < r < t + \varepsilon.$$

It follows that $\mathbf{X}_{0,t_n}[i] = \mathbf{X}_{0,t}[i]$ for all n large enough such that $t - \varepsilon < t_n < t + \varepsilon$. Since $x_n \to x$, Lemma 4.20 now tells us that $\mathbf{X}_{0,t}(x_n)(i) = \mathbf{X}_{0,t}(x)(i)$ for all n large enough. ∎

Proof of Theorem 4.18 By Lemma 4.19, almost surely, for each $s \in \mathbb{R}$ and $x \in S^\Lambda$, the evolution equation (4.10) has a unique solution $(X_t^{s,x})_{t \geq s}$ with initial state $X_s^{s,x} = x$, which allows us to define random maps $(\mathbf{X}_{s,t})_{s \leq t}$ satisfying (4.13) as in (4.12). By Lemma 4.20 the maps $\mathbf{X}_{s,t} : S^\Lambda \to S^\Lambda$ are continuous. To see that (4.14) defines a Feller semigroup, we need to check that

(i) $(x,t) \mapsto P_t(x, \cdot\,)$ is a continuous map from $E \times [0,\infty)$ into $\mathcal{M}_1(E)$,
(ii) $P_0 = 1$ and $P_s P_t = P_{s+t} \quad (s,t \geq 0)$.

Property (i) follows from Lemma 4.21 and the fact that a.s. convergence implies weak convergence in law. To prove (ii), we observe that for each bounded measurable $f : S^\Lambda \to \mathbb{R}$,

$$P_0 f(x) = \mathbb{E}\big[f(\mathbf{X}_{0,0}(x))\big] = f(x)$$

and, for every measurable $A \subset S^\Lambda$,

$$\begin{aligned} P_{s+t}(x,A) &= \mathbb{P}\big[\mathbf{X}_{s,s+t} \circ \mathbf{X}_{0,s}(x) \in A\big] \\ &= \int \mathbb{P}\big[\mathbf{X}_{0,s}(x) \in \mathrm{d}y\big]\, \mathbb{P}\big[\mathbf{X}_{s,s+t}(y) \in A \,|\, \mathbf{X}_{0,s}(x) = y\big] \\ &= \int \mathbb{P}\big[\mathbf{X}_{0,s}(x) \in \mathrm{d}y\big]\, \mathbb{P}\big[\mathbf{X}_{s,s+t}(y) \in A\big] \\ &= \int P_s(x,\mathrm{d}y) P_t(y,A) = P_s P_t(x,A), \end{aligned}$$

where we have used the stationarity and independent increments of the stochastic flow.

Using the fact that the stochastic flow $(\mathbf{X}_{s,t})_{s \leq t}$ is stationary and has independent increments, it is now straightforward to check that the process in (4.15) is distributed as a Markov process with semigroup $(P_t)_{t \geq 0}$. Note that $(X_t)_{t \geq 0}$ has cadlag sample paths since solutions to the evolution equation (4.10) are cadlag. This completes the proof. ∎

4.4 Finite Approximation and Perturbations

In this section, we use the methods of the previous section to prove two additional results that are sometimes useful. In (4.12), we defined a stochastic flow $(\mathbf{X}_{s,t})_{s \leq t}$ in terms of a graphical representation ω by means of the evolution equation (4.10). Our first aim in this section is to show that $(\mathbf{X}_{s,t})_{s \leq t}$ can alternatively be defined by approximating ω with finite sets.

We first need a simple fact about continuous-time Markov chains. Let S be a countable set. Recall that a random mapping representation of the generator G of a continuous-time Markov chain with state space S is an expression of the form (2.21) where the rates $(r_m)_{m \in \mathcal{G}}$ satisfy (2.20). Let ω be the graphical representation associated with such a random mapping representation and let ω' be a finite subset of ω. Then, for each $s \le u$, we define a map $\mathbf{X}^{\omega'}_{s,u}: S \to S$ by

$$\begin{aligned} &\mathbf{X}^{\omega'}_{s,u} := m_n \circ \cdots \circ m_1, \quad \text{where} \\ &\big\{(m,t) \in \omega' : t \in (s,u]\big\} = \big\{(m_1,t_1),\ldots,(m_n,t_n)\big\} \quad \text{with} \quad t_1 < \cdots < t_n. \end{aligned} \tag{4.27}$$

In words, this says that $\mathbf{X}^{\omega'}_{s,u}$ is the concatenation of the maps m for which $(m,t) \in \omega'$ with $t \in (s,u]$, ordered by their times. Below, we equip S with the discrete topology, so (4.28) simply says that $\mathbf{X}^{\omega_n}_{s,u}(x) = \mathbf{X}_{s,u}(x)$ for all n large enough.

Lemma 4.22 (Finite approximation of Markov chains) *Let G be the generator of a nonexplosive continuous-time Markov chain with countable state space S, let ω be the graphical representation associated with a random mapping representation of G, and let $(\mathbf{X}_{s,t})_{s\le t}$ be the stochastic flow defined in terms of ω. Then almost surely, for all $s \le u$, for all $x \in S$, and for each sequence ω_n of finite subsets of ω such that $\omega_n \uparrow \omega$, one has that*[9]

$$\mathbf{X}^{\omega_n}_{s,u}(x) \underset{n\to\infty}{\longrightarrow} \mathbf{X}_{s,u}(x). \tag{4.28}$$

An analogue statement holds for the backward stochastic flow defined in terms of ω as in Theorem 2.24, where now $\mathbf{X}^{\omega_n}_{u,s}$ is defined by ordering the elements of $\{(m,t) \in \omega_n : t \in (s,u]\}$ in decreasing order of time.

Proof Fix $s \le u$ and $x \in S$. Then $[s,u] \ni t \mapsto \mathbf{X}_{s,t}(x)$ is piecewise constant and right-continuous. We set

$$\omega^* := \big\{(m,t) \in \omega : t \in (s,u],\ m\big(\mathbf{X}_{s,t-}(x)\big) \ne \mathbf{X}_{s,t-}(x)\big\},$$

and write

$$\omega^* = \big\{(m_1,t_1),\ldots,(m_k,t_k)\big\} \quad \text{with} \quad t_1 < \cdots < t_k.$$

Then

$$\mathbf{X}_{s,u}(x) = m_k \circ \cdots \circ m_1(x). \tag{4.29}$$

It follows from the definition of ω^* that if $(m,t) \in \omega\backslash\omega^*$ for some $t \in (s,u]$, then $m\big(\mathbf{X}_{s,t-}(x)\big) = \mathbf{X}_{s,t-}(x)$. This implies that the right-hand side of (4.29) does not change if in the concatenation of maps we add additional maps from $\omega\backslash\omega^*$ at their appropriate times. It follows that for all n large enough such that $\omega_n \supset \omega^*$, one has

$$\mathbf{X}^{\omega_n}_{s,u}(x) = m_k \circ \cdots \circ m_1(x) = \mathbf{X}_{s,u}(x).$$

∎

If ω is the graphical representation of an interacting particle system and ω' is a finite subset of ω, then we define $\mathbf{X}^{\omega'}_{s,t}$ in the same way as in (4.27).

[9] For those who like abstract mathematics, another way to formulate (4.28) is as follows. Denoting by F the set of finite subsets of ω, equipped with the partial order of set inclusion, we have that $(\mathbf{X}^{\omega}_{s,u})^{\omega\in F}$ is a *net* in the topological meaning of the word, and (4.28) says that this net converges pointwise to $\mathbf{X}_{s,u}$.

Proposition 4.23 (Finite approximation) *Under the assumptions of Theorem 4.18, the stochastic flow* $(\mathbf{X}_{s,t})_{s\leq t}$ *has the property that almost surely, for all* $s \leq u$*, for all* $x \in S^\Lambda$*, and for each sequence* ω_n *of finite subsets of* ω *such that* $\omega_n \uparrow \omega$*, one has that*

$$\mathbf{X}^{\omega_n}_{s,u}(x) \underset{n\to\infty}{\longrightarrow} \mathbf{X}_{s,u}(x) \tag{4.30}$$

with respect to the product topology on S^Λ.

Proof We continue to use the notation $\mathbf{X}_{s,u}[i](x) := \mathbf{X}_{s,u}(x)(i)$ $(i \in \Lambda,\ x \in S^\Lambda)$ and, as we did before, define $\phi_i \in C(S^\Lambda, S)$ by $\phi_i(x) := x(i)$ $(i \in \Lambda)$. Fix $s \leq u$ and $i \in \Lambda$. Then (4.25) tells us that

$$\mathbf{X}_{s,u}[i] = \mathbf{F}_{u,s}(\phi_i).$$

Similarly, for each finite $\omega' \subset \omega$, if $(m_1, t_t), \ldots, (m_k, t_k)$ are the elements of $\{(m,t) \in \omega' : t \in (s,u]\}$, ordered according to their times, then

$$\mathbf{X}^{\omega'}_{s,u}[i] = \phi_i \circ m_k \circ \cdots \circ m_1 = \mathbf{F}^{\omega'}_{u,s}(\phi_i).$$

Applying Lemma 4.22 to the backward stochastic flow $(\mathbf{F}_{u,s})_{u\geq s}$ we see that if $\omega_n \uparrow \omega$, then for all n large enough

$$\mathbf{X}^{\omega_n}_{s,u}[i] = \mathbf{F}^{\omega_n}_{u,s}(\phi_i) = \mathbf{F}_{u,s}(\phi_i) = \mathbf{X}_{s,u}[i].$$

Since this holds for each $i \in \Lambda$, it follows that $\mathbf{X}^{\omega_n}_{s,u}(x)$ converges to $\mathbf{X}_{s,u}(x)$ in the product topology for each $x \in S^\Lambda$. ∎

We have already noticed the similarity between condition (2.41) of Theorem 2.35 and condition (4.11) of Theorem 4.14, the only difference being that $\mathcal{R}^\uparrow_i(m)$ in (2.41) (ii) is replaced by $\mathcal{R}^\downarrow_i(m)$ in (4.11) (ii). In Theorem 2.35, we assumed that $m(\underline{0}) = \underline{0}$ for all $m \in \mathcal{G}$, which implies that the all-zero configuration is a trivial fixed point of the evolution equation (4.10). The theorem then showed that under the condition (2.41), finite perturbations of the all-zero initial state have finite consequences at later times. We show here that this statement generalizes to arbitrary initial states.

Proposition 4.24 (Finite perturbations) *Assume that the conditions of Theorem 4.18 are satisfied and that (2.41) holds, that is,*

$$\text{(i)}\ \sup_{i\in\Lambda}\sum_{m\in\mathcal{G}} r_m 1_{\mathcal{D}(m)}(i) < \infty, \qquad \text{(ii)}\ \sup_{i\in\Lambda}\sum_{m\in\mathcal{G}} r_m\big|\mathcal{R}^\uparrow_i(m)\backslash\{i\}\big| < \infty. \tag{4.31}$$

Then almost surely, for each $x, y \in S^\Lambda$ *such that* $A := \{i \in \Lambda : x(i) \neq y(i)\}$ *is finite, one has that*

$$A_{s,t} := \big\{i \in \Lambda : \mathbf{X}_{s,t}(x)(i) \neq \mathbf{X}_{s,t}(y)(i)\big\} \tag{4.32}$$

is finite for all $s \leq t$*. If* s, t, x, y *are deterministic, then*

$$\mathbb{E}\big[|A_{s,t}|\big] \leq |A|e^{K_\uparrow(t-s)}, \tag{4.33}$$

where $K_\uparrow$ *is defined in (4.16).*

Proof We assume without loss of generality that $|S| \geq 2$. Indeed, if $|S| = 1$, then $A_{s,t} = \emptyset$ for all s, t, x, y and the statements are trivial. It follows from (4.31) that if $|\mathcal{R}_i^{\uparrow}(m)| = \infty$ for some $i \in \Lambda$ and $m \in \mathcal{G}$, then $r_m = 0$, so we can without loss of generality assume that $\mathcal{R}_i^{\uparrow}(m)$ is finite for all $i \in \Lambda$ and $m \in \mathcal{G}$. Let $\mathcal{P}$ denote the set of all subsets of Λ, and set $\mathcal{P}_{\text{fin}} := \{A \in \mathcal{P} : |A| < \infty\}$. For each $m \in \mathcal{G}$, we define a map $\overline{m}\colon \mathcal{P}_{\text{fin}} \to \mathcal{P}_{\text{fin}}$ by

$$\overline{m}(A) := \bigcup_{i \in A} \mathcal{R}_i^{\uparrow}(m) \qquad (A \in \mathcal{P}_{\text{fin}}).$$

We claim that setting

$$\overline{G} f(A) := \sum_{m \in \mathcal{G}} r_m \big\{ f\big(\overline{m}(A)\big) - f(A) \big\}$$

defines the generator of a nonexplosive continuous-time Markov chain $(Z_t)_{t \geq 0}$ with state space $\mathcal{P}_{\text{fin}}$, and that this Markov chain satisfies

$$\mathbb{E}^A\big[|Z_t|\big] \leq |A| e^{K_\uparrow t} \qquad (t \geq 0,\ A \in \mathcal{P}_{\text{fin}}). \tag{4.34}$$

We start by checking (2.20), which in the present context says that

$$\sum_{m:\, \overline{m}(A) \neq A} r_m < \infty \quad \forall A \in \mathcal{P}_{\text{fin}}.$$

If $\overline{m}(A) \neq A$, then either there exists an $i \in A$ such that $i \notin \overline{m}(A)$, or there exists a $j \notin A$ such that $j \in \overline{m}(A)$. In the first case, $i \notin \mathcal{R}_i^{\uparrow}(m)$, which by the fact that $|S| \geq 2$ implies $i \in \mathcal{D}(m)$. In the second case there must be an $i \in A$ such that $j \in \mathcal{R}_i^{\uparrow}(m)$. It follows that

$$\begin{aligned}\sum_{m:\, \overline{m}(A) \neq A} r_m &\leq \sum_{m \in \mathcal{G}} \sum_{i \in A} 1_{\mathcal{D}(m)}(i) r_m + \sum_{m \in \mathcal{G}} \sum_{i \in A} \sum_{j \in \mathcal{R}_i^{\uparrow}(m) \setminus A} r_m \\ &\leq |A| \sup_{i \in \Lambda} \sum_{m \in \mathcal{G}} 1_{\mathcal{D}(m)}(i) r_m + |A| \sup_{i \in \Lambda} \sum_{m \in \mathcal{G}} \big|\mathcal{R}_i^{\uparrow}(m) \setminus \{i\}\big| r_m,\end{aligned} \tag{4.35}$$

which is finite by (4.31). Lemma 2.21 now implies that $\overline{G}$ is the generator of a (possibly explosive) continuous-time Markov chain with state space $\mathcal{P}_{\text{fin}}$. To prove nonexplosiveness and the exponential bound (4.34), we apply Theorem 2.12 to the Lyapunov function $L(A) := |A|$. Formula (4.35) shows that condition (i) of Theorem 2.12 is satisfied, so it remains to check that $\overline{G} L \leq K_\uparrow L$. This is very similar to the last steps of the proof of Proposition 4.16. Indeed,

$$\big|\overline{m}(A)\big| - \big|A\big| \leq \sum_{i \in A} \big(|\mathcal{R}_i^{\uparrow}(m)| - 1\big),$$

which implies that

$$\overline{G} L(A) \leq \sum_{m \in \mathcal{G}} \sum_{i \in A} r_m \big(|\mathcal{R}_i^{\uparrow}(m)| - 1\big) \leq K_\uparrow |A| = K_\uparrow L(A).$$

Let $(\mathbf{Z}_{s,t})_{s \leq t}$ be the stochastic flow constructed from the Poisson point set $\overline{\omega} := \{(\overline{m}, t) : (m, t) \in \omega\}$. Since $\overline{G}$ is nonexplosive, Theorem 2.22 tells us that almost surely, $\mathbf{Z}_{s,t}$ maps $\mathcal{P}_{\text{fin}}$ into itself for each $s \leq t$. To complete the proof, we will show that almost surely, for each $s \leq t$ and $x, y \in S^\Lambda$ such that $A := \{i \in \Lambda : x(i) \neq y(i)\}$ is finite, the set $A_{s,t}$ defined in (4.32) is contained in $\mathbf{Z}_{s,t}(A)$. The basic observation we need is that for each $m \in \mathcal{G}$ and $x, y \in S^\Lambda$ such that $\{i \in \Lambda : x(i) \neq y(i)\}$ is finite

$$\big\{i \in \Lambda : x(i) \neq y(i)\big\} \subset A \quad \text{implies} \quad \big\{i \in \Lambda : m(x)(i) \neq m(y)(i)\big\} \subset \overline{m}(A).$$

Using this, we see that for each $x, y \in S^\Lambda$ such that $\{i \in \Lambda : x(i) \neq y(i)\}$ is finite, for each $s \leq u$, and for each finite $\omega_n \subset \omega$,

$$\big\{i \in \Lambda : x(i) \neq y(i)\big\} \subset A \quad \text{implies} \quad \big\{i \in \Lambda : \mathbf{X}^{\omega_n}_{s,u}(x)(i) \neq \mathbf{X}^{\omega_n}_{s,u}(y)(i)\big\} \subset \mathbf{Z}^{\omega_n}_{s,u}(A).$$

The claim now follows by letting $\omega_n \uparrow \omega$, using Lemma 4.22 and Proposition 4.23. ∎

4.5 Generator Construction

Although Theorem 4.14 gives us an explicit way to construct the Feller semigroup associated with an interacting particle system, it does not tell us very much about its generator. To fill this gap, we need a bit more theory. For any continuous function $f : S^\Lambda \to \mathbb{R}$ and $i \in \Lambda$, we define

$$\delta f(i) := \sup\big\{|f(x) - f(y)| : x, y \in S^\Lambda,\ x(j) = y(j)\ \forall j \neq i\big\}.$$

Note that $\delta f(i)$ measures how much $f(x)$ can change if we change x only in the point i. We call δf the *variation* of f.[10]

Lemma 4.25 (Variation of a function) *Let $f \in C(S^\Lambda)$. Then*

$$\big|f(x) - f(y)\big| \leq \sum_{i : x(i) \neq y(i)} \delta f(i) \qquad \big(f \in C(S^\Lambda),\ x, y \in S^\Lambda\big). \tag{4.36}$$

Proof Let n be the number of sites i where x and y differ. Enumerate these sites as $\{i : x(i) \neq y(i)\} = \{i_1, \ldots, i_n\}$ or $= \{i_1, i_2, \ldots\}$ depending on whether n is finite or not. For $0 \leq k < n + 1$, set

$$z_k(i) := \begin{cases} y(i) & \text{if } i \in \{i_1, \ldots, i_k\}, \\ x(i) & \text{otherwise.} \end{cases}$$

If n is finite, then

$$\big|f(x) - f(y)\big| \leq \sum_{k=1}^{n} \big|f(z_k) - f(z_{k-1})\big| \leq \sum_{k=1}^{n} \delta f(i_k)$$

and we are done. If n is infinite, then the same argument gives

$$\big|f(x) - f(z_m)\big| \leq \sum_{k=1}^{m} \delta f(i_k) \qquad (m \geq 1).$$

[10] This definition is similar to, but different from, the more usual definition of the (total) variation of a function of one real variable. With functions of one real variable, the total variation is the maximal sum of all changes in the value of the function as one gradually increases the real variable. For functions on S^Λ, the idea is similar, but instead of increasing a real variable, we will gradually change a configuration x by modifying its coordinates one by one.

Since $z_m \to y$ pointwise and f is continuous, (4.36) now follows by letting $m \to \infty$. ∎

We define spaces of functions by

$$C_{\text{sum}} = C_{\text{sum}}(S^\Lambda) := \big\{f \in C(S^\Lambda) : \sum_i \delta f(i) < \infty\big\},$$
$$C_{\text{fin}} = C_{\text{fin}}(S^\Lambda) := \big\{f \in C(S^\Lambda) : \delta f(i) = 0 \text{ for all but finitely many } i\big\}.$$

We say that functions in C_{sum} are of *summable variation*. The next exercise shows that functions in C_{fin} depend on finitely many coordinates only.

Exercise 4.26 *Let us say that a function $f\colon S^\Lambda \to \mathbb{R}$* depends on finitely many coordinates *if there exists a finite set $A \subset \Lambda$ and a function $f'\colon S^A \to \mathbb{R}$ such that*

$$f\big((x(i))_{i\in\Lambda}\big) = f'\big((x(i))_{i\in A}\big) \qquad \big(x \in S^\Lambda\big).$$

Show that each function that depends on finitely many coordinates is continuous, that

$$C_{\text{fin}}(S^\Lambda) = \big\{f \in C(S^\Lambda) : f \text{ depends on finitely many coordinates}\big\},$$

and that $C_{\text{fin}}(S^\Lambda)$ is a dense linear subspace of the Banach space $C(S^\Lambda)$ of all continuous real functions on S^Λ, equipped with the supremum norm.

Exercise 4.27 *Define $f\colon \{0,1\}^{\mathbb{N}} \to \mathbb{R}$ by*

$$f(x) := \frac{1}{1+r} \quad \textit{with} \quad r := \inf\{i \geq 0 : x(i) = 1\}.$$

Show that $f \in C(\{0,1\}^{\mathbb{N}})$ but $f \notin C_{\text{sum}}(\{0,1\}^{\mathbb{N}})$.

In what follows, we assume that $\mathcal{G}$ is a countable collection of continuous maps $m\colon S^\Lambda \to S^\Lambda$ and that $(r_m)_{m\in\mathcal{G}}$ are nonnegative rates.

Lemma 4.28 (Domain of pregenerator) *Assume (4.11) (i) and let K_0 be the constant defined in (4.16). Then, for each $f \in C_{\text{sum}}(S^\Lambda)$,*

$$\sum_{m\in\mathcal{G}} r_m \big|f(m(x)) - f(x)\big| \leq K_0 \sum_{i\in\Lambda} \delta f(i).$$

In particular, for each $f \in C_{\text{sum}}(S^\Lambda)$ and $x \in S^\Lambda$, the right-hand side of (4.1) is absolutely summable and Gf is well-defined.

Proof This follows by writing

$$\sum_{m\in\mathcal{G}} r_m\big|f(m(x)) - f(x)\big| \leq \sum_{m\in\mathcal{G}} r_m \sum_{i\in\mathcal{D}(m)} \delta f(i) = \sum_{i\in\Lambda} \delta f(i) \sum_{\substack{m\in\mathcal{G}\\ \mathcal{D}(m)\ni i}} r_m \leq K_0 \sum_{i\in\Lambda} \delta f(i).$$

∎

The following theorem is the main result of the present section.

Theorem 4.29 (Generator construction of particle systems) *Assume that the rates $(r_m)_{m\in\mathcal{G}}$ satisfy (4.11), let $(P_t)_{t\geq 0}$ be the Feller semigroup defined in (4.14), and let G be the linear operator with domain $\mathcal{D}(G) := C_{\text{sum}}$ defined by (4.1). Then G is closable, and its closure $\overline{G}$ is the generator of $(P_t)_{t\geq 0}$. Moreover, C_{fin} is a core for G.*

To prepare for the proof of Theorem 4.29, we need a few lemmas.

Lemma 4.30 (Generator on local functions) *Under the assumptions of Theorem 4.29, one has* $\lim_{t\downarrow 0} t^{-1}(P_t f - f) = Gf$ *for all* $f \in \mathcal{C}_{\mathrm{fin}}$*, where the limit holds with respect to the topology on* $\mathcal{C}(S^\Lambda)$.

Proof Since $f \in \mathcal{C}_{\mathrm{fin}}$, by Exercise 4.26, there exists some finite $A \subset \Lambda$ such that f depends only on the coordinates in A. Let $\mathcal{G}^A := \{m \in \mathcal{G} : \mathcal{D}(m) \cap A \neq \emptyset\}$ denote the set of maps $m \in \mathcal{G}$ that can potentially change the state in A. We introduce the notation

$$\omega_{s,t} := \{(m,r) \in \omega : r \in (s,t]\} \qquad (s \le t)$$

and we let $\omega^A_{s,t}$ denote the set of Poisson points $(m,r) \in \omega_{s,t}$ with $m \in \mathcal{G}^A$. If $\omega^A_{0,t} = \emptyset$, then $f(\mathbf{X}_{0,t}(x)) = f(x)$. Also, if $\omega^A_{0,t}$ contains a single element (m,s), then $f(\mathbf{X}_{0,t}(x)) = f(m(x))$. Therefore,

$$\begin{aligned} P_t f(x) = &\mathbb{E}\big[f\big(\mathbf{X}_{0,t}(x)\big)\big] = f(x)\mathbb{P}[\omega^A_{0,t} = \emptyset] \\ &+ \sum_{m\in\mathcal{G}^A} f\big(m(x)\big)\mathbb{P}\big[\omega^A_{0,t} = \{(m,s)\} \text{ for some } 0 < s \le t\big] \\ &+ \mathbb{E}\big[f\big(\mathbf{X}_{0,t}(x)\big)1_{\{|\omega^A_{0,t}| \ge 2\}}\big]. \end{aligned}$$

Here, setting $R := \sum_{m\in\mathcal{G}^A} r_m$, which is finite by the finiteness of A and (4.11) (i), we have

$$\begin{aligned} &\mathbb{P}[\omega^A_{0,t} = \emptyset] = e^{-Rt}, \\ &\mathbb{P}\big[\omega^A_{0,t} = \{(m,s)\} \text{ for some } 0 < s \le t\big] = t r_m e^{-Rt} \quad (m \in \mathcal{G}^A), \\ &\mathbb{P}[|\omega^A_{0,t}| \ge 2] = 1 - e^{-Rt} - tRe^{-Rt}. \end{aligned}$$

It follows that

$$\begin{aligned} P_t f(x) &= e^{-Rt} f(x) + te^{-Rt} \sum_{m\in\mathcal{G}^A} r_m f\big(m(x)\big) + O_x(t^2), \\ &= f(x) + t \sum_{m\in\mathcal{G}^A} r_m\big\{f\big(m(x)\big) - f\big(x\big)\big\} + O_x(t^2), \end{aligned} \tag{4.37}$$

where $O_x(t^2)$ denotes a function, which may differ from line to line, that has the property that $\limsup_{t\to\infty} t^{-2} \sup_{x\in S^\Lambda} |O_x(t^2)| < \infty$. Indeed, in the first line of (4.37),

$$O_x(t^2) = \mathbb{E}\big[f\big(\mathbf{X}_{0,t}(x)\big)1_{\{|\omega^A_{0,t}| \ge 2\}}\big].$$

This can be estimated as

$$\sup_{x\in S^\Lambda} |O_x(t^2)| \le \|f\|_\infty \mathbb{P}[|\omega^A_{0,t}| \ge 2] = \|f\|_\infty\big(1 - e^{-Rt} - tRe^{-Rt}\big),$$

which using the fact that $e^{-Rt} = 1 - Rt + O(t^2)$ as $t \to 0$ shows that the error term in the first line of (4.37) is of order t^2 uniformly in $x \in S^\Lambda$. The second line of (4.37) now also follows readily, with a somewhat different definition of $O_x(t^2)$.

Since $f(m(x)) = f(x)$ if $m \notin \mathcal{G}^A$, formula (4.37) implies that

$$t^{-1}\big(P_t f(x) - f(x)\big) = Gf(x) + O_x(t),$$

where $O_x(t)$ denotes a term that is of order t as $t \to 0$, uniformly in $x \in S^\Lambda$. This shows that

$$\lim_{t\to\infty} \big\|t^{-1}(P_t f - f) - Gf\big\|_\infty = 0,$$

as claimed. ∎

Lemma 4.31 (Approximation by local functions) *Assume (4.11) (i). Then for all $f \in \mathcal{C}_{\rm sum}$, there exist $f_n \in \mathcal{C}_{\rm fin}$ such that $\|f_n - f\|_\infty \longrightarrow 0$ and $\|Gf_n - Gf\|_\infty \longrightarrow 0$.*

Proof Choose finite $\Lambda_n \uparrow \Lambda$, set $\Gamma_n := \Lambda\backslash\Lambda_n$, fix $z \in S^\Lambda$, and for each $x \in S^\Lambda$ define $x^n \longrightarrow x$ by

$$x^n(i) := \begin{cases} x(i) & \text{if } i \in \Lambda_n, \\ z(i) & \text{if } i \in \Gamma_n. \end{cases}$$

Fix $f \in \mathcal{C}_{\rm sum}$ and define $f_n(x) := f(x^n)$ $(x \in S^\Lambda)$. Then f_n depends only on the coordinates in Λ_n; hence, $f_n \in \mathcal{C}_{\rm fin}$. If $x_n, x \in S^\Lambda$ satisfy $x_n \longrightarrow x$, then by the continuity of f, we have $f_n(x_n) = f(x_n^n) \longrightarrow f$, so applying Lemma 4.1, we see that

$$\|f_n - f\|_\infty \underset{n\to\infty}{\longrightarrow} 0.$$

To prove that also $\|Gf_n - Gf\|_\infty \longrightarrow 0$, we observe that

$$\begin{aligned} &|Gf_n(x) - Gf(x)| \\ &\quad = \Big|\sum_{m\in\mathcal{G}} r_m\big(f_n(m(x)) - f_n(x)\big) - \sum_{m\in\mathcal{G}} r_m\big(f(m(x)) - f(x)\big)\Big| \\ &\quad \le \sum_{m\in\mathcal{G}} r_m\big|f(m(x)^n) - f(x^n) - f(m(x)) + f(x)\big|. \end{aligned} \tag{4.38}$$

On the one hand, we have

$$\begin{aligned} &\big|f(m(x)^n) - f(x^n) - f(m(x)) + f(x)\big| \\ &\quad \le \big|f(m(x)^n) - f(x^n)\big| + \big|f(m(x)) - f(x)\big| \le 2\sum_{i\in\mathcal{D}(m)} \delta f(i), \end{aligned}$$

while on the other hand, we can estimate the same quantity as

$$\le \big|f(m(x)^n) - f(m(x))\big| + \big|f(x^n) - f(x)\big| \le 2\sum_{i\in\Gamma_n} \delta f(i).$$

Let $A \subset \Lambda$ be finite. Inserting either of our two estimates into (4.38), depending on whether $\mathcal{D}(m) \cap A \neq \emptyset$ or not, we find that

$$\begin{aligned} \|Gf_n - Gf\|_\infty &\le 2\sum_{\substack{m\in\mathcal{G}\\ \mathcal{D}(m)\cap A\neq\emptyset}} r_m \sum_{i\in\Gamma_n} \delta f(i) + 2\sum_{\substack{m\in\mathcal{G}\\ \mathcal{D}(m)\cap A=\emptyset}} r_m \sum_{i\in\mathcal{D}(m)} \delta f(i) \\ &\le 2K_0|A|\sum_{i\in\Gamma_n} \delta f(i) + 2\sum_{i\in\Lambda} \delta f(i) \sum_{\substack{m\in\mathcal{G}\\ \mathcal{D}(m)\cap A=\emptyset\\ \mathcal{D}(m)\ni i}} r_m, \end{aligned}$$

where K_0 is the constant defined in (4.16). It follows that

$$\limsup_{n\to\infty} \|Gf_n - Gf\|_\infty \le 2\sum_{i\in\Lambda\backslash A} \delta f(i) \sum_{\substack{m\in\mathcal{G}\\ \mathcal{D}(m)\ni i}} r_m \le 2K_0 \sum_{i\in\Lambda\backslash A} \delta f(i).$$

Since A is arbitrary, letting $A \uparrow \Lambda$, we see that $\limsup_n \|Gf_n - Gf\|_\infty = 0$. ∎

Lemma 4.32 (Functions of summable variation) *Under the assumptions of Theorem 4.29, one has*

$$\sum_{i\in\Lambda}\delta P_t f(i)\le e^{K_\downarrow t}\sum_{i\in\Lambda}\delta f(i)\qquad \big(t\ge 0,\ f\in C_{\rm sum}(S^\Lambda)\big),$$

where $K_\downarrow$ is the constant from (4.16). In particular, for each $t\ge 0$, P_t maps $C_{\rm sum}(S^\Lambda)$ into itself.

Proof Fix $i\in\Lambda$. As we have done before, for $j\in\Lambda$ we define $\phi_j\in C(S^\Lambda,S)$ by $\phi_j(x):=x(j)$ $(x\in S^\Lambda)$. Then for each $x,y\in S^\Lambda$ such that $x(k)=y(k)$ for all $k\neq i$, we can estimate using (4.36) and (4.25)

$$\begin{aligned}
|P_tf(x)-P_tf(y)| &= \big|\mathbb{E}[f(\mathbf{X}_{0,t}(x))]-\mathbb{E}[f(\mathbf{X}_{0,t}(y))]\big|\\
&\le \mathbb{E}\big[|f(\mathbf{X}_{0,t}(x))-f(\mathbf{X}_{0,t}(y))|\big]\le \mathbb{E}\big[\textstyle\sum_{j:\mathbf{X}_{0,t}(x)(j)\neq\mathbf{X}_{0,t}(y)(j)}\delta f(j)\big]\\
&=\sum_j\mathbb{P}\big[\mathbf{X}_{0,t}(x)(j)\neq\mathbf{X}_{0,t}(y)(j)\big]\delta f(j)=\sum_j\mathbb{P}\big[\mathbf{F}_{t,0}(\phi_j)(x)\neq\mathbf{F}_{t,0}(\phi_j)(y)\big]\delta f(j)\\
&\le\sum_j\mathbb{P}\big[i\in\mathcal{R}(\mathbf{F}_{t,0}(\phi_j))\big]\delta f(j).
\end{aligned}$$

By formula (4.18) of Proposition 4.16, it follows that

$$\begin{aligned}
\sum_i\delta P_tf(i)&\le\sum_{ij}\mathbb{P}\big[i\in\mathcal{R}(\mathbf{F}_{t,0}(\phi_j))\big]\delta f(j)\\
&=\sum_j\mathbb{E}\big[|\mathcal{R}(\mathbf{F}_{t,0}(\phi_j))|\big]\delta f(j)\le e^{K_\downarrow t}\sum_j\delta f(j).
\end{aligned}$$

∎

Proof of Theorem 4.29 Let H be the full generator of $(P_t)_{t\ge0}$ and let $\mathcal{D}(H)$ denote its domain. Then Lemma 4.30 shows that $C_{\rm fin}\subset\mathcal{D}(H)$ and $Gf=Hf$ for all $f\in C_{\rm fin}$. By Lemma 4.31, it follows that $C_{\rm sum}\subset\mathcal{D}(H)$ and $Gf=Hf$ for all $f\in C_{\rm sum}$. To complete the proof, it suffices to show that $C_{\rm fin}$, and hence also the larger $C_{\rm sum}$, is a core for H.

We first prove that $C_{\rm sum}$ is a core for H. We will apply Lemma 4.8. We will show that for each $r>K_\downarrow$, where $K_\downarrow$ is the constant from (4.16), and for each $f\in C_{\rm sum}(S^\Lambda)$, there exists a $p_r\in C_{\rm sum}(S^\Lambda)$ that solves the Laplace equation $(r-G)p_r=f$. Since $C_{\rm sum}(S^\Lambda)$ is dense in $C(S^\Lambda)$ by Exercise 4.26, the claim then follows from the equivalence of (i) and (ii) of Lemma 4.8.

Fix $r>K_\downarrow$ and $f\in C_{\rm sum}(S^\Lambda)$. We need to find a $p_r\in C_{\rm sum}(S^\Lambda)$ that solves the Laplace equation $(r-G)p_r=f$. In the light of Lemma 4.6 a natural candidate for such a function is

$$p_r:=\int_0^\infty e^{-rt}P_tf\,\mathrm{d}t$$

and we will show that this p_r indeed satisfies $p_r\in C_{\rm sum}(S^\Lambda)$ and $(r-G)p_r=f$. It follows from Theorem 4.9 that $p_r\in\mathcal{D}(H)$ and $(r-H)p_r=f$. Thus, it suffices to show that $p_r\in C_{\rm sum}$. To see this, note that if $x(j)=y(j)$ for all $j\neq i$, then

$$\begin{aligned}
|p_r(x)-p_r(y)|&=\Big|\int_0^\infty e^{-rt}P_tf(x)\,\mathrm{d}t-\int_0^\infty e^{-rt}P_tf(y)\mathrm{d}t\Big|\\
&\le\int_0^\infty e^{-rt}\big|P_tf(x)-P_tf(y)\big|\,\mathrm{d}t\le\int_0^\infty e^{-rt}\delta P_tf(i)\,\mathrm{d}t,
\end{aligned}$$

and therefore, by Lemma 4.32, and our assumption that $r > K_\downarrow$

$$\sum_i \delta p(i) \leq \int_0^\infty e^{-rt} \sum_i \delta P_t f(i)\, \mathrm{d}t \leq \Big(\sum_i \delta f(i)\Big) \int_0^\infty e^{-rt} e^{K_\downarrow t}\, \mathrm{d}t < \infty,$$

which proves that $p_r \in C_{\mathrm{sum}}$. This completes the proof that C_{sum} is a core for H, that is, the closure of $G|_{C_{\mathrm{sum}}}$ is H. By Lemma 4.31, the closure of $G|_{C_{\mathrm{fin}}}$ contains $G|_{C_{\mathrm{sum}}}$, so we see that C_{fin} is also a core for H. ∎

The following lemma is sometimes useful.

Lemma 4.33 (Differentiation of semigroup) *Assume that the rates $(r_m)_{m\in\mathcal{G}}$ satisfy (4.11), let $(P_t)_{t\geq 0}$ be the Feller semigroup defined in (4.14), and let G be the linear operator with domain $\mathcal{D}(G) := C_{\mathrm{sum}}(S^\Lambda)$ defined by (4.1). Then, for each $f \in C_{\mathrm{sum}}(S^\Lambda)$, $t \mapsto P_t f$ is a continuously differentiable function from $[0,\infty)$ to $C(S^\Lambda)$ satisfying $P_0 f = f$, $P_t f \in C_{\mathrm{sum}}(S^\Lambda)$, and $\frac{\partial}{\partial t} P_t f = G P_t f = P_t G f$ for each $t \geq 0$.*

Proof The statement that $\frac{\partial}{\partial t} P_t f = P_t G f$ holds for any Feller semigroup and f in the domain of its generator; see [EK86, Prop 1.1.5]. The remaining statements are a direct consequence of Proposition 4.5, Lemma 4.32, and Theorem 4.29. A direct proof based on our definition of $(P_t)_{t\geq 0}$ (not using Hille–Yosida theory) is also possible, but quite long and technical. ∎

We conclude this section by proving an analogue of Proposition 2.3 for interacting particle systems. We continue to assume that S is a finite set, Λ is countable, $\mathcal{G}$ is a collection of continuous maps $m\colon S^\Lambda \to S^\Lambda$, and $(r_m)_{m\in\mathcal{G}}$ are nonnegative rates satisfying (4.11). By Theorem 4.29, the linear operator G with domain $\mathcal{D}(G) := C_{\mathrm{sum}}$ defined in (4.1) is closable, and its closure $\overline{G}$ generates a Feller semigroup $(P_t)_{t\geq 0}$. We need the following simple lemma.

Lemma 4.34 (Closedness under multiplication) *For each $f, g \in C_{\mathrm{sum}}$, the pointwise product fg is an element of C_{sum}.*

Proof For each $i \in \Lambda$ and $x, y \in S^\Lambda$ such that $x(j) = y(j)$ for all $j \neq i$, we can estimate

$$\begin{aligned}\big|f(x)g(x) - f(y)g(y)\big| &\leq \big|f(x)g(x) - f(y)g(x)\big| + \big|f(y)g(x) - f(y)g(y)\big| \\ &\leq \|g\|_\infty \cdot \big|f(x) - f(y)\big| + \|f\|_\infty \big|g(x) - g(y)\big|,\end{aligned}$$

which tells us that

$$\delta(fg)(i) \leq \|g\|_\infty \delta f(i) + \|f\|_\infty \delta g(i).$$

Summing over i yields the claim. ∎

Lemma 4.34 allows us to define $\Gamma_G\colon C_{\mathrm{sum}} \times C_{\mathrm{sum}} \to C$ by

$$\Gamma_G(f,g) := G(fg) - (Gf)g - f(Gg) \qquad (f, g \in C_{\mathrm{sum}}).$$

A calculation similar to the one below (2.10) shows that

$$\Gamma_G(f,g)(x) = \sum_{m\in\mathcal{G}} r_m \big\{f\big(m(x)\big) - f\big(x\big)\big\}\big\{g\big(m(x)\big) - g\big(x\big)\big\}. \tag{4.39}$$

The following proposition generalizes Proposition 2.3 to interacting particle systems.

Proposition 4.35 (Covariance formula) *Assume that the rates $(r_m)_{m\in\mathcal{G}}$ satisfy (4.11), let $(P_t)_{t\geq 0}$ be the Feller semigroup defined in (4.14), and let G be the linear operator with domain $\mathcal{D}(G) := \mathcal{C}_{\rm sum}$ defined in (4.1). Then for each probability measure μ on S^Λ, one has*

$$\mathrm{Cov}_{\mu P_t}(f,g) = \mathrm{Cov}_\mu(P_t f, P_t g) + \int_0^t \mathrm{d}s\, \mu P_{t-s}\Gamma_G(P_s f, P_s g) \quad (f,g \in \mathcal{C}_{\rm sum}).$$

Proof The proof is essentially the same as in the finite case (Proposition 2.3). Using Lemmas 4.33 and 4.34, and the continuity of P_{t_1}, we obtain that

$$\begin{aligned}
\tfrac{\partial}{\partial t_1} P_{t_1}\big((P_{t_2}f)(P_{t_3}g)\big) &= P_{t_1} G\big((P_{t_2}f)(P_{t_3}g)\big),\\
\tfrac{\partial}{\partial t_2} P_{t_1}\big((P_{t_2}f)(P_{t_3}g)\big) &= P_{t_1}\big((GP_{t_2}f)(P_{t_3}g)\big),\\
\tfrac{\partial}{\partial t_3} P_{t_1}\big((P_{t_2}f)(P_{t_3}g)\big) &= P_{t_1}\big((P_{t_2}f)(GP_{t_3}g)\big).
\end{aligned}$$

The rest of the proof is the same. ∎

Some Bibliographical Remarks

In 1972, several authors published results of various degrees of generality showing that interacting particle systems on infinite lattices are well-defined. Harris [Har72] used the Poisson approach. His result applies only to finite range interactions on $\mathbb{Z}^d$. Instead of using the backward in time process, he argued forward in time, using percolation theory to show that if t is small enough, then the lattice can randomly be divided into finite pieces that mutually do not interact with each other during the time interval $(0,t]$.

Liggett [Lig72], on the other hand, gave a direct proof that the closure of G generates a Feller semigroup $(P_t)_{t\geq 0}$, and then invoked the abstract result Theorem 4.2 about Feller processes to prove the existence of a corresponding Markov process with cadlag sample paths. This result is more widely applicable than Harris' result and made it to Liggett's famous book [Lig85, Theorem I.3.9]. Liggett's conditions are similar to condition (4.11) of our Theorem 4.14, but there are also some differences. Liggett does not write his generators in terms of local maps but in terms of local probability kernels. This way of writing the generator is more general and sometimes (e.g., for stochastic Ising models) more natural than our approach using local maps. It is worth noting that Liggett's construction, like ours, depends on a clever way of writing the generator that is in general not unique.

Liggett's book [Lig85] does not treat graphical representations in the generality of our Theorem 4.14, but he does use explicit Poisson constructions for some specific systems, such as the contact process. He does not actually prove that these Poisson constructions yield the same process as the generator construction but apparently finds this self-evident. (Equivalence of the two constructions follows from our Theorem 4.29 but alternatively can also be proved by approximation with finite systems, using approximation results such as [Lig85, Cor. I.3.14].)

Liggett's [Lig85, Theorem I.3.9] allows for the case that the local state space S is a (not necessarily finite) compact metrizable space. This is occasionally convenient. For example, this allows one to construct voter models with infinitely many types, where at time zero, the types $(X_0(i))_{i\in\Lambda}$ are i.i.d. and uniformly distributed on $S = [0,1]$. We have made essential use of the finiteness of S in several places. For example, the state space $\mathcal{C}(S^\Lambda, T)$ of the backward in time process is no longer countable if S is not finite and, as explained in the text preceding Exercise 4.15, solutions to the evolution equation (4.10) may no longer be unique if S is allowed to be a general compact metrizable space. With some extra work, these difficulties can presumably be overcome (e.g., by requiring that solutions to (4.10) are cadlag with respect to the discrete topology on S) but for simplicity we restrict ourselves to finite local state spaces. Non-compact local state spaces are more tricky, see [Lig85, Chapter IX]. An alternative treatment of non-compact local state spaces that works only for processes with finite range interactions is given in [Pen08].

The backward in time process of Proposition 4.16 will come back in Chapter 6 when we discuss duality of interacting particle systems. It is also interesting to look at the mean-field limit of this process. One can show that in the mean-field limit, the process

$$\big(\mathcal{R}(\Phi_t)\big)_{t\geq 0} \quad \text{with} \quad \Phi_t := \mathbf{F}_{u,u-t}(\Phi_0) \quad (t \geq 0)$$

behaves as a branching process. As a result, solutions to the mean-field equation can be represented in terms of a stochastic process on the genealogical tree of a branching process. This is explained in [MSS20].

4.6 Ergodicity

The proofs of Theorems 4.14 and 4.29 were quite long. Luckily, they yield more information than just the fact that the interacting particle systems we are interested in are well-defined. The basic phenomenon that motivates the study of interacting particle systems is collective behavior. The general picture is that for weak strengths of the interaction, different parts of space behave essentially independently, but for sufficiently strong interaction, it may happen that all sites start to coordinate their behavior, giving rise to multiple invariant laws or even more exotic phenomena such as periodic behavior.

As a result of the methods of the previous sections, we will be able to prove results that confirm the "easy" part of this picture, namely the absence of collective behavior for weak strengths of the interaction.

If X is a Markov process with state space E and transition probabilities $(P_t)_{t\geq 0}$, then by definition, an *invariant law* of X is a probability measure ν on E such that

$$\nu P_t = \nu \qquad (t \geq 0).$$

This says that if we start the process in the initial law $\mathbb{P}[X_0 \in \cdot\,] = \nu$, then $\mathbb{P}[X_t \in \cdot\,] = \nu$ for all $t \geq 0$. As a consequence, one can construct a stationary process $(X_t)_{t\in\mathbb{R}}$ such that (compare (4.2))

$$\mathbb{P}\big[X_u \in \cdot \,\big|\, (X_s)_{-\infty<s\leq t}\big] = P_{u-t}(X_t, \cdot\,) \quad \text{a.s.} \qquad (t \leq u), \tag{4.40}$$

and $\mathbb{P}[X_t \in \cdot\,] = \nu$ for all $t \in \mathbb{R}$. Conversely, the existence of such a stationary Markov process implies that the law at any time $\nu := \mathbb{P}[X_t \in \cdot\,]$ must be an invariant law.

Theorem 4.36 (Ergodicity) *Let X be an interacting particle system with state space of the form S^Λ and generator G of the form (4.1), and assume that the rates $(r_m)_{m\in\mathcal{G}}$ satisfy (4.11). Let T be a finite set with at least two elements.*

(a) *Assume that the constant* $K_\downarrow$ *from (4.16) satisfies* $K_\downarrow < 0$. *Then the backward in time process satisfies*

$$\lim_{t\to\infty} \big|\mathcal{R}(\mathbf{F}_{u,u-t}(\phi)\big| = 0 \quad \text{a.s.} \quad \big(u \in \mathbb{R},\ \phi \in C(S^\Lambda, T)\big). \tag{4.41}$$

(b) *Assume that the backward in time process satisfies (4.41). Then the interacting particle system* X *has a unique invariant law* ν, *and*

$$\mathbb{P}^x\big[X_t \in \cdot\,\big] \underset{t\to\infty}{\Longrightarrow} \nu \qquad (x \in S^\Lambda). \tag{4.42}$$

Moreover, there exists an a.s. unique cadlag process $(X_t)_{t\in\mathbb{R}}$ *such that*

$$X_t = \mathfrak{m}_t^\omega(X_{t-}) \qquad (t \in \mathbb{R}), \tag{4.43}$$

and $(X_t)_{t\in\mathbb{R}}$ *is distributed as the stationary Markov process corresponding to the invariant law* ν.

Proof Part (a) is immediate from formula (4.18) of Proposition 4.16. Let Ψ_i denote the set of functions $\phi\colon S^\Lambda \to T$ that depend only on $x(i)$. Note that this set is finite. By (4.25) and the assumption that T has at least two elements

$$\mathcal{R}\big(\mathbf{X}_{s,t}[i]\big) \subset \bigcup_{\phi\in\Psi_i} \mathcal{R}\big(\mathbf{F}_{t,s}(\phi)\big) \quad (s \le t,\ i \in \Lambda),$$

so (4.41) implies that

$$\lim_{s\to-\infty} \big|\mathcal{R}(\mathbf{X}_{s,t}[i])\big| = 0 \quad \text{a.s.} \quad \big(t \in \mathbb{R},\ i \in \Lambda\big). \tag{4.44}$$

It follows from the definition of $(\mathbf{X}_{s,t})_{s\le t}$ in (4.12) that the function $t \mapsto \mathbf{X}_{s,t}[i]$ jumps only at times for which there exists a $(m,t) \in \omega$ such that $i \in \mathcal{D}(m)$. Since this set is locally finite by (4.11), we can replace the order of the "almost sure" and "for all t" statements; that is, (4.44) holds almost surely for all t simultaneously. Formula (4.44) says that for low enough s, the function $\mathbf{X}_{s,t}[i]$ is constant, which implies that for each $z \in S^\Lambda$ the a.s. limit

$$X_t(i) := \lim_{s\to-\infty} \mathbf{X}_{s,t}(z)(i) \qquad (i \in \Lambda,\ t \in \mathbb{R}) \tag{4.45}$$

exists and does not depend on the choice of the configuration $z \in S^\Lambda$. Using the continuity of $\mathbf{X}_{s,u}$ (which is proved in Theorem 4.14) and the flow property, we see that a.s.

$$\mathbf{X}_{t,u}(X_t) = \lim_{s\to-\infty} \mathbf{X}_{t,u} \circ \mathbf{X}_{s,t}(z) = X_u \qquad (t \le u),$$

which implies that $(X_t)_{t\in\mathbb{R}}$ solves (4.43). If $(X'_t)_{t\in\mathbb{R}}$ is another solution, then for all s low enough so that $\mathcal{R}(\mathbf{X}_{s,t}[i]) = \emptyset$,

$$X'_t(i) = \mathbf{X}_{s,t}(X'_s)(i) = \mathbf{X}_{s,t}(z)(i) = X_t(i),$$

which shows that solutions to (4.43) are unique.

We claim that $X = (X_t)_{t\in\mathbb{R}}$ is Markov with respect to the transition probabilities $(P_t)_{t\ge0}$ in the sense of (4.40). Indeed, for almost every trajectory $(x_s)_{-\infty<s\le t}$ with respect to the law

of $(X_s)_{-\infty<s\leq t}$, we have

$$\begin{aligned}
&\mathbb{P}\big[X_u \in \cdot \,\big|\, (X_s)_{-\infty<s\leq t} = (x_s)_{-\infty<s\leq t}\big] \\
&\quad = \mathbb{P}\big[\lim_{s\to-\infty} \mathbf{X}_{t,u} \circ \mathbf{X}_{s,t}(z) \in \cdot \,\big|\, (X_s)_{-\infty<s\leq t} = (x_s)_{-\infty<s\leq t}\big] \\
&\quad \stackrel{1}{=} \mathbb{P}\big[\mathbf{X}_{t,u}(X_t) \in \cdot \,\big|\, (X_s)_{-\infty<s\leq t} = (x_s)_{-\infty<s\leq t}\big] \\
&\quad \stackrel{2}{=} \mathbb{P}\big[\mathbf{X}_{t,u}(x_t) \in \cdot \,\big|\, (X_s)_{-\infty<s\leq t} = (x_s)_{-\infty<s\leq t}\big] \\
&\quad \stackrel{3}{=} \mathbb{P}\big[\mathbf{X}_{t,u}(x_t) \in \cdot\,\big] = P_{u-t}(x_t, \cdot),
\end{aligned}$$

where in step 1 we have used the continuity of the map $\mathbf{X}_{t,u}$, in step 2 we have replaced X_t by x_t, and in step 3 we have used that the random variables $\mathbf{X}_{t,u}$ and $(X_s)_{-\infty<s\leq t}$ are independent, since they are functions of the restriction of the Poisson set ω to the disjoint sets $\mathcal{G} \times (t, u]$ and $\mathcal{G} \times (-\infty, t]$, respectively. By the stationarity of the stochastic flow,

$$\nu := \mathbb{P}\big[X_t \in \cdot\,\big] \qquad (t \in \mathbb{R})$$

does not depend on $t \in \mathbb{R}$, and since X is Markov, this defines an invariant law ν. Since

$$\mathbb{P}^x\big[X_t \in \cdot\,\big] = \mathbb{P}\big[\mathbf{X}_{-t,0}(x) \in \cdot\,\big]$$

and since by (4.45), we have

$$\mathbf{X}_{-t,0}(x) \underset{t\to\infty}{\longrightarrow} X_0 \quad \text{a.s.} \qquad (x \in S^\Lambda)$$

with respect to the topology of pointwise convergence, we conclude that (4.42) holds. ∎

Remark It is possible for an interacting particle system to be ergodic in the sense of (4.42), while (4.41) does not hold. In such a situation, it is not clear if solutions to (4.43) are a.s. unique. Even if there are multiple invariant laws, one can ask if (4.43) has an a.s. unique solution subject to the condition that $(X_t)_{t\in\mathbb{R}}$ is stationary with a given invariant law. Not much is known about this, but these questions are related to the concept of endogeny of recursive tree processes [AB05, MSS20].

We note that (4.42) says that if we start the process in an arbitrary initial state x, then the law at time t converges weakly[11] as $t \to \infty$ to the invariant law ν. This property is often described by saying that the interacting particle system is *ergodic*. Indeed, this implies that the corresponding stationary process $(X_t)_{t\in\mathbb{R}}$ is ergodic in the usual sense of that word, that is, the σ-field of events that are invariant under translations in time is trivial. The converse conclusion cannot be drawn, however, so the traditional way of describing (4.42) as "ergodicity" is a bit of a bad habit.

We have split Theorem 4.36 into parts (a) and (b) since the condition (4.41) is sometimes satisfied even when the constant $K_\downarrow$ from (4.16) is positive. Indeed, we will later see that for the contact process, the condition (4.41) is sharp, but the condition $K_\downarrow < 0$ is not. In Exercise 5.21, we will calculate the constant $K_\downarrow$ for the contact process and deduce that this process is ergodic for small values of the infection rate.

Theorem 4.36 is similar, but not identical, to [Lig85, Thm I.4.1]. For Theorem 4.36 (a) and (b) to be applicable, one needs to be able to express the generator in terms of local maps such

[11] Here weak convergence is of course w.r.t. our topology on S^Λ, that is, w.r.t. the product topology.

that the constant $K_{\downarrow}$ from (4.16) is negative. For [Lig85, Thm I.4.1], one needs to express the generator in a convenient way in terms of local transition kernels. For certain problems, the latter approach is more natural, and [Lig85, Thm I.4.1] yields sharper estimates for the regime where ergodicity holds.

4.7 Application to the Ising Model

The Ising model with Glauber dynamics has been introduced in Section 1.4. So far, we have not shown how to represent the generator of this interacting particle system in terms of local maps. In the present section, we will fill this gap. We will only consider the ferromagnetic case $\beta \geq 0$. As an application of the theory developed so far, we will then show that the Ising model with Glauber dynamics is well-defined for all values of its parameter and ergodic for β sufficiently small. Our construction will also prepare for the next chapter, where we discuss monotone interacting particle systems, by showing that the Ising model with Glauber dynamics can be represented in monotone maps.

We recall from Section 1.4 that the Ising model with Glauber dynamics on a graph (Λ, E) is the interacting particle system with state space $\{-1,+1\}^{\Lambda}$ and dynamics such that

$$\text{site } i \text{ flips to the value } \sigma \text{ with rate} \quad r_i^{\sigma}(x) := \frac{e^{\beta N_{x,i}(\sigma)}}{e^{\beta N_{x,i}(+1)} + e^{\beta N_{x,i}(-1)}},$$

where

$$N_{x,i}(\sigma) := \sum_{j \in \mathcal{N}_i} 1_{\{x(j) = \sigma\}} \qquad (\sigma \in \{-1,+1\})$$

denotes the number of neighbors of i that have the spin value σ. For each $i \in \Lambda$, let K_i^{β} denote the probability kernel on $\{-1,+1\}^{\Lambda}$ defined as

$$K_i^{\beta}(x,y) := \begin{cases} r_i^{\sigma}(x) & \text{if } y = m_i^{\sigma}(x) \quad (\sigma \in \{-1,+1\}), \\ 0 & \text{otherwise,} \end{cases}$$

where $m_i^{\sigma}(x)$, defined in (1.14), denotes the configuration x with the spin at i flipped to σ. Then the generator (1.13) of the Ising model takes the form

$$G_{\text{Ising}} f = \sum_{i \in \Lambda} \{K_i^{\beta} f - f\}, \tag{4.46}$$

which is an expression of the form (2.7) but not a random mapping representation of the form (4.1). To find a random mapping representation for G_{Ising} in terms of local maps as in (4.1), it suffices to find a random mapping representation for the kernels K_i^{β}. This needs some preparation. Let

$$M_{x,i} := N_{x,i}(+) - N_{x,i}(-) = \sum_{j \in \mathcal{N}_i} x(j)$$

denote the *local magnetization* in the neighborhood $\mathcal{N}_i$ of i. Since $N_{x,i}(+) + N_{x,i}(-) = |\mathcal{N}_i|$, we can rewrite the probability under K_i^β of flipping to the spin value +1 as

$$r_i^+(x) = \frac{e^{\beta N_{x,i}(+1)}}{e^{\beta N_{x,i}(+1)} + e^{\beta N_{x,i}(-1)}} = \frac{e^{\beta(|\mathcal{N}_i|+M_{x,i})/2}}{e^{\beta(|\mathcal{N}_i|+M_{x,i})/2} + e^{\beta(|\mathcal{N}_i|-M_{x,i})/2}}$$

$$= \frac{e^{\frac{1}{2}\beta M_{x,i}}}{e^{\frac{1}{2}\beta M_{x,i}} + e^{-\frac{1}{2}\beta M_{x,i}}} = \tfrac{1}{2}\Big(1 + \frac{e^{\frac{1}{2}\beta M_{x,i}} - e^{-\frac{1}{2}\beta M_{x,i}}}{e^{\frac{1}{2}\beta M_{x,i}} + e^{-\frac{1}{2}\beta M_{x,i}}}\Big)$$

$$= \tfrac{1}{2}\big(1 + \tanh(\tfrac{1}{2}\beta M_{x,i})\big).$$

Similarly, the probability of flipping to -1 is $r_i^-(x) = \frac{1}{2}(1 - \tanh(\frac{1}{2}\beta M_{x,i})) = 1 - r_i^+(x)$.

For (mainly notational) simplicity, let us assume that each site i has the same number of neighbors in the graph (Λ, E), so that the size of the neighborhood

$$N := |\mathcal{N}_i| \qquad (i \in \Lambda)$$

does not depend on $i \in \Lambda$. Then $M_{x,i}$ takes values in $\{-N, -N+2, \dots, N\}$. We observe that for $\beta > 0$ the function $z \mapsto \frac{1}{2}(1 + \tanh(\frac{1}{2}\beta z))$ is increasing (see Figure 4.2). Inspired by this, for $L = -N-1, -N+1, \dots, N+1$, we define local maps $m_{i,L}$ by

$$m_{i,L}(x)(j) := \begin{cases} +1 & \text{if } j = i \text{ and } M_{x,i} > L, \\ -1 & \text{if } j = i \text{ and } M_{x,i} < L, \\ x(j) & \text{if } j \neq i. \end{cases} \tag{4.47}$$

We try a generator of the form

$$G_{\text{Ising}} f(x) = \sum_{i \in \Lambda} \sum_{L=-N-1}^{N+1} r_{i,L} \big\{ f\big(m_{i,L}(x)\big) - f\big(x\big) \big\}, \tag{4.48}$$

where we sum only over odd L, and the constants $r_{i,L} \geq 0$ are probabilities that need to be chosen in such a way that

$$K_i^\beta(x, y) = \sum_{L=-N-1}^{N+1} r_{i,L} 1_{\{m_{i,L}(x) = y\}} \tag{4.49}$$

is a random mapping representation of the kernel K_i^β. Consider x, y such that $x(i) = -1$, $y(i) = +1$, and $x(j) = y(j)$ for all $j \neq i$. For such x, y, (4.49) yields the equation

$$\tfrac{1}{2}\big(1 + \tanh(\tfrac{1}{2}\beta M_{x,i})\big) = r_i^+(x) = \sum_{L=-N-1}^{M_{x,i}-1} r_{i,L}$$

Similarly, for x, y such that $x(i) = +1$, $y(i) = -1$, and $x(j) = y(j)$ for all $j \neq i$, formula (4.49) yields

$$\tfrac{1}{2}\big(1 - \tanh(\tfrac{1}{2}\beta M_{x,i})\big) = r_i^-(x) = \sum_{L=M_{x,i}+1}^{N+1} r_{i,L}.$$

From this, we see that (4.49) is satisfied for (see Figure 4.2)

$$r_{i,L} := \begin{cases} \frac{1}{2}\big(1 + \tanh(-\frac{1}{2}\beta N)\big) & \text{if } L = -N-1, \\ \frac{1}{2}\tanh(\frac{1}{2}\beta(L+1)) - \frac{1}{2}\tanh(\frac{1}{2}\beta(L-1)) & \text{if } -N+1 \leq L \leq N-1, \\ \frac{1}{2}\big(1 - \tanh(\frac{1}{2}\beta N)\big) & \text{if } L = N+1, \end{cases} \tag{4.50}$$

which has the effect that the generator in (4.48) equals the one in (4.46). We observe that this is even true for $\beta = 0$: in this case, $r_{i,-N-1} = \frac{1}{2} = r_{i,N+1}$, and all other probabilities are zero.

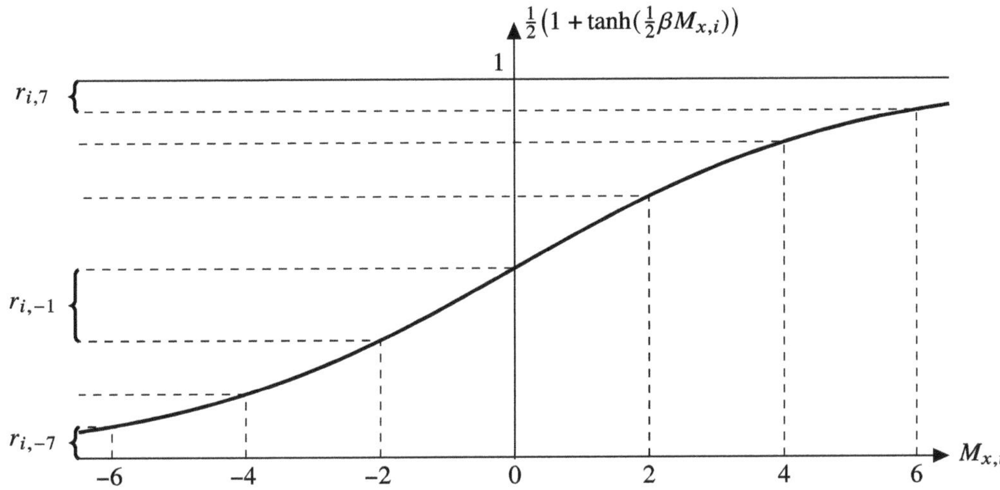

Figure 4.2 Definition of the rates $r_{i,L}$ from (4.50). In this example, $N = 6$ and $\beta = 0.4$.

Theorem 4.37 (Existence and ergodicity of the Ising model) *Consider an Ising model with Glauber dynamics on a countable graph Λ in which each lattice point i has exactly $|\mathcal{N}_i| = N \geq 2$ neighbors, that is, the Markov process X with state space $\{-1,+1\}^\Lambda$ and generator $G_{\rm Ising}$ given by (4.46). Then, for each $\beta \geq 0$, the closure of $G_{\rm Ising}$ generates a Feller semigroup. Moreover, for each*

$$0 \leq \beta < N^{-1}\big(\log(N+1) - \log(N-1)\big), \tag{4.51}$$

the Markov process with generator $\overline{G}_{\rm Ising}$ has a unique invariant law ν, and the process started in an arbitrary initial state x satisfies

$$\mathbb{P}^x\big[X_t \in \cdot\,\big] \underset{t\to\infty}{\Longrightarrow} \nu \qquad \big(x \in \{-1,+1\}^\Lambda\big).$$

For the critical point of the Ising model on $\mathbb{Z}^2$, Theorem 4.37 yields the bound

$$0.1277 \approx \tfrac{1}{4}(\log 5 - \log 3) \leq \beta_{\rm c},$$

which should be compared with the known value $\beta_{\rm c} = \log(1+\sqrt{2}) \approx 0.8814$ from (1.16).

Proof of Theorem 4.37 We use the representation (4.48). We observe that $\mathcal{D}(m_{i,L}) = \{i\}$ is the set of lattice points whose spin value can be changed by the map $m_{i,L}$. The set of lattice points that are $m_{i,L}$-relevant for i is given by

$$\mathcal{R}_i^{\downarrow}(m_{i,L}) = \begin{cases} \emptyset & \text{if } L = -N-1 \quad \text{or} \quad L = N+1, \\ \mathcal{N}_i & \text{otherwise.} \end{cases}$$

Here we have used that $-N-1 < M_{x,i} \leq N+1$ holds always, so $m_{-N-1}(x)(i) = +1$ and $m_{N+1}(x)(i) = -1$ regardless of what x is. On the other hand, in all other cases, the value of each lattice point $j \in \mathcal{N}_i$ can potentially make a difference for the outcome $m_{i,L}(x)(i)$.

By Theorem 4.29, to conclude that the closure of G_{Ising} generates a Feller semigroup, it suffices to check that the expressions in (4.11) (i) and (ii) are finite. In our case, these expressions are

$$\text{(i)} \sup_{i\in\Lambda} \sum_{L=-N-1}^{N+1} r_{i,L} = 1 \qquad \text{(ii)} \sup_{i\in\Lambda} \sum_{L=-N+1}^{N-1} r_{i,L}|\mathcal{N}_i| \leq N.$$

To prove ergodicity for β small enough, we apply Theorem 4.36. We calculate the constant $K_{\downarrow}$ from (4.16).

$$\begin{aligned} K_{\downarrow} &= \sum_{L=-N-1}^{N+1} r_{i,L}\big(|\mathcal{R}_i^{\downarrow}(m_{i,L})| - 1\big) = -r_{i,-N-1} - r_{i,N+1} + \sum_{L=-N+1}^{N-1} r_{i,L}(N-1) \\ &= -1 + N \sum_{L=-N+1}^{N-1} r_{i,L} = N\tanh(\tfrac{1}{2}\beta N) - 1. \end{aligned}$$

This is negative if and only if

$$\begin{aligned} &N\frac{e^{\frac{1}{2}\beta N} - e^{-\frac{1}{2}\beta N}}{e^{\frac{1}{2}\beta N} + e^{-\frac{1}{2}\beta N}} < 1 \quad \Leftrightarrow \quad N\big(e^{\frac{1}{2}\beta N} - e^{-\frac{1}{2}\beta N}\big) < e^{\frac{1}{2}\beta N} + e^{-\frac{1}{2}\beta N} \\ &\Leftrightarrow \quad (N-1)e^{\frac{1}{2}\beta N} < (N+1)e^{-\frac{1}{2}\beta N} \quad \Leftrightarrow \quad e^{\beta N} < \frac{N+1}{N-1}, \end{aligned}$$

which yields condition (4.51). ∎

Exercise 4.38 *Show that the antiferromagnetic Ising model is ergodic if*

$$-N^{-1}\big(\log(N+1) - \log(N-1)\big) < \beta \leq 0.$$

4.8 Further Results

In the present section, we collect a number of technical results of a general nature that will be needed in later chapters. On a first reading, readers are advised to skip the present section and refer back to specific results when the need arises. The only result of the present section that is perhaps of some intrinsic value is Theorem 4.43, which together with Corollary 4.44, implies that the transition probabilities of interacting particle systems on infinite lattices can be approximated by those on finite lattices, something that we have been using implicitly when doing simulations. An alternative way to see this is to use Proposition 4.23 that we have already proved.

Let E be a compact metrizable space. By definition, a collection of functions $\mathcal{H} \subset C(E)$ is *distribution determining* if for probability measures μ, ν on E

$$\mu f = \nu f \ \forall f \in \mathcal{H} \quad \text{implies} \quad \mu = \nu.$$

We say that $\mathcal{H}$ *separates points* if for all $x, y \in E$ such that $x \neq y$, there exists an $f \in \mathcal{H}$ such that $f(x) \neq f(y)$. We say that $\mathcal{H}$ is *closed under products* if $f, g \in \mathcal{H}$ implies $fg \in \mathcal{H}$.

Lemma 4.39 (Application of Stone–Weierstrass) *Let E be a compact metrizable space. Assume that $\mathcal{H} \subset C(E)$ separates points and is closed under products. Then $\mathcal{H}$ is distribution determining.*

Proof If $\mu f = \nu f$ for all $f \in \mathcal{H}$, then we can add the constant function 1 to $\mathcal{H}$ and retain this property. In a next step, we can add all linear combinations of functions in $\mathcal{H}$ to the set $\mathcal{H}$; by the linearity of the integral, it will then still be true that $\mu f = \nu f$ for all $f \in \mathcal{H}$. But now $\mathcal{H}$ is an algebra that separates points and vanishes nowhere, so by the Stone–Weierstrass theorem, $\mathcal{H}$ is dense in $C(E)$. If $f_n \in \mathcal{H}$, $f \in C(E)$, and $\|f_n - f\|_\infty \to 0$, then $\mu f_n \to \mu f$ and likewise for ν, so we conclude that $\mu f = \nu f$ for all $f \in C(E)$. If $A \subset E$ is a closed set, then the function $f(x) := d(x, A)$ is continuous, where d is a metric generating the topology on E and $d(x, A) := \inf_{y \in A} d(x, y)$ denotes the distance of x to A. Now the functions $f_n := 1 \wedge nf$ are also continuous and $f_n \uparrow 1_{A^c}$, so by the continuity of the integral with respect to increasing sequences, we see that $\mu(O) = \nu(O)$ for every open set $O \subset E$. Since the open sets are closed under intersections, it follows that $\mu(A) = \nu(A)$ for every element A of the σ-algebra generated by the open sets, that is, the Borel σ-field $\mathcal{B}(E)$. ∎

Lemma 4.40 (Weak convergence) *Let E be a compact metrizable space. Assume that $\mu_n \in \mathcal{M}_1(E)$ have the property that $\lim_{n\to\infty} \mu_n f$ exists for all $f \in \mathcal{H}$, where $\mathcal{H} \subset C(E)$ is distribution determining. Then there exists a $\mu \in \mathcal{M}_1(E)$ such that $\mu_n \Rightarrow \mu$.*

Proof By Prohorov's theorem, the space $\mathcal{M}_1(E)$, equipped with the topology of weak convergence, is compact. Therefore, to prove the statement, it suffices to show that the sequence μ_n has not more than one cluster point, that is, it suffices to show that if μ, μ' are subsequential limits, then $\mu' = \mu$. Clearly, μ, μ' must satisfy $\mu' f = \mu f$ for all $f \in \mathcal{H}$, so the claim follows from the assumption that $\mathcal{H}$ is distribution determining. ∎

Lemma 4.41 (Continuous probability kernels) *Let E be a compact metrizable space, and let K be a continuous probability kernel on E. Then, for any $\mu_n, \mu \in \mathcal{M}_1(E)$ and $f_n, f \in C(E)$,*

$$\begin{aligned} &\mu_n \underset{n\to\infty}{\Longrightarrow} \mu \quad \textit{implies} \quad \mu_n K \underset{n\to\infty}{\Longrightarrow} \mu K \\ \textit{and} \quad &\|f_n - f\|_\infty \underset{n\to\infty}{\longrightarrow} 0 \quad \textit{implies} \quad \|K f_n - K f\|_\infty \underset{n\to\infty}{\longrightarrow} 0. \end{aligned}$$

Proof Since K is a continuous probability kernel, its associated operator maps the space $C(E)$ into itself, so $\mu_n \Rightarrow \mu$ implies that $\mu_n(Kf) \Rightarrow \mu(Kf)$ for all $f \in C(E)$, or equivalently $(\mu_n K)f \Rightarrow (\mu K)f$ for all $f \in C(E)$, that is, the measures $\mu_n K$ converge weakly to μK.

The second statement follows from the linearity and monotonicity of K and the fact that $K1 = 1$, which together imply that $\|K f_n - K f\|_\infty \leq \|f_n - f\|_\infty$. ∎

Lemma 4.42 (Long-time limits) *Let E be a compact metrizable space, and let $(P_t)_{t\geq 0}$ be the transition probabilities of a Feller process in E. Let $\mu \in \mathcal{M}_1(E)$ and assume that*

$$\mu P_t \underset{t\to\infty}{\Longrightarrow} \nu$$

for some $\nu \in \mathcal{M}_1(E)$. Then ν is an invariant law of the Feller process with transition probabilities $(P_t)_{t\geq 0}$.

Proof Using Lemma 4.41, this follows by writing

$$\nu P_t = (\lim_{s\to\infty} \mu P_s) P_t = \lim_{s\to\infty} \mu P_s P_t = \lim_{s\to\infty} \mu P_{s+t} = \nu.$$

∎

The following theorem follows from [Kal97, Thm 17.25], where it is moreover shown that the condition (4.52) implies convergence in distribution of the associated Feller processes, viewed as random variables taking values in the space $\mathcal{D}_E[0,\infty)$ of cadlag paths with values in E. Note that in (4.52), $\to$ (of course) means convergence in the topology we have defined on $\mathcal{C}(E)$, that is, convergence w.r.t. the supremum norm.

Theorem 4.43 (Limits of semigroups) *Let E be a compact metrizable space, and let G_n, G be generators of Feller processes in E. Assume that there exists a linear operator A on $\mathcal{C}(E)$ such that $\overline{A} = G$ and*

$$\forall f \in \mathcal{D}(A)\ \exists f_n \in \mathcal{D}(G_n) \text{ such that } f_n \to f \quad \text{and} \quad G_n f_n \to Af. \tag{4.52}$$

Then the Feller semigroups $(P^n_t)_{t\geq 0}$ and $(P_t)_{t\geq 0}$ with generators G_n and G, respectively, satisfy

$$\sup_{t\in[0,T]} \|P^n_t f - P_t f\|_\infty \underset{n\to\infty}{\longrightarrow} 0 \quad \big(f \in \mathcal{C}(E),\ T < \infty\big).$$

Moreover, if $\mu_n, \mu \in \mathcal{M}_1(E)$, then

$$\mu_n \underset{n\to\infty}{\Longrightarrow} \mu \quad \text{implies} \quad \mu_n P^n_t \underset{n\to\infty}{\Longrightarrow} \mu P_t \qquad (t \geq 0).$$

We note that in the case of interacting particle systems, Theorem 4.29 implies the following.

Corollary 4.44 (Convergence of particle systems) *Let S be a finite set and let Λ be countable. Let G_n, G be generators of interacting particle systems in S^Λ and assume that G_n, G can be written in the form (4.1) with rates satisfying (4.11). Assume moreover that*

$$\|G_n f - Gf\|_\infty \underset{n\to\infty}{\longrightarrow} 0 \qquad \big(f \in \mathcal{C}_{\rm fin}(S^\Lambda)\big).$$

Then the generators G_n, G satisfy (4.52) with A the restriction of G to $\mathcal{C}_{\rm fin}(S^\Lambda)$.

Theorem 4.43 has the following useful consequence.

Proposition 4.45 (Limits of invariant laws) *Let E be a compact metrizable space, and let G_n, G be generators of Feller processes in E satisfying (4.52). Let $\nu_n, \nu \in \mathcal{M}_1(E)$ and assume that for each n, the measure ν_n is an invariant law of the Feller process with generator G_n. Then $\nu_n \Rightarrow \nu$ implies that ν is an invariant law of the Feller process with generator G.*

Proof Using Theorem 4.43, this follows simply by observing that

$$\nu P_t = \lim_{n\to\infty} \nu_n P_t^n = \lim_{n\to\infty} \nu_n = \nu$$

for each $t \geq 0$. ∎

5

Monotonicity

5.1 The Stochastic Order

If the local state space S of an interacting particle system is partially ordered, then we equip the product space S^Λ with the *product order*

$$x \leq y \quad \text{if and only if} \quad x(i) \leq y(i) \ \forall i \in \Lambda.$$

Many well-known interacting particle systems use the local state space $S = \{0, 1\}$, which is of course equipped with a natural order $0 \leq 1$. Often, it is useful to prove comparison results, that say that two interacting particle systems X and Y can be coupled in such a way that $X_t \leq Y_t$ for all $t \geq 0$. Here X and Y may be different systems, started in the same initial state, or also two copies of the same interacting particle system, started in initial states such that $X_0 \leq Y_0$. A useful tool in such comparison arguments is the stochastic order, which is the subject of the present section. We will come back to interacting particle systems in the next section.

We recall that if S and T are partially ordered sets, then a function $f: S \to T$ is called *monotone* if and only if $x \leq y$ implies $f(x) \leq f(y)$. In particular, this definition also applies to real-valued functions (where we equip $\mathbb{R}$ with the well-known order). Throughout this section, E is a compact metrizable space that is equipped with a partial order $\leq$ that is *compatible with the topology* in the sense that

$$\big\{(x, y) \in E^2 : x \leq y\big\} \text{ is closed in the product topology on } E^2.$$

We recall that $B(E)$ and $C(E)$ denote the spaces of Borel measurable bounded functions and continuous functions $f: E \to \mathbb{R}$, respectively. We set

$$B^+(E) := \big\{f \in B(E) : f \text{ is monotone}\big\} \quad \text{and} \quad C^+(E) := B^+(E) \cap C(E).$$

We need the following technical result.

Lemma 5.1 (Distribution determining property) *If μ, ν are probability measures on E such that $\int \mu(\mathrm{d}x) f(x) = \int \nu(\mathrm{d}x) f(x)$ for all $f \in C^+(E)$, then $\mu = \nu$.*

Proof Let $\mathcal{F} := \{f \in C^+(\mathcal{X}) : f \geq 0\}$. By Lemma 4.39, it suffices to show that $\mathcal{F}$ is closed under products in the sense that $f, g \in \mathcal{F}$ imply $fg \in \mathcal{F}$, and separates points in the sense that for each $x, y \in \mathcal{X}$ with $x \neq y$, there exists an $f \in \mathcal{F}$ such that $f(x) \neq f(y)$. Closedness under products is trivial. Showing that $\mathcal{F}$ separates points is considerably more work. Assume that $x, y \in E$ satisfy $x \neq y$. Then either $x \not\leq y$ or $y \not\leq x$. By symmetry we may assume that we are in the second case. Let $E_0 := \{z \in E : z \leq x\}$ and $E_1 := \{z \in E : z \geq y\}$. Then

$E_0 \cap E_1 = \emptyset$. Since the partial order is compatible with the topology, E_0 and E_1 are closed subsets of E. Our assumptions on E imply that it is a "compact ordered space" as defined in [Nac65, Section 1.4], which by the corollary to [Nac65, Theorem I.3.4] implies that E is a "normally ordered space" as defined in [Nac65, Section 1.2]. We can then apply [Nac65, Theorem I.2.1] to conclude that there exists a continuous monotone function $f: E \to [0,1]$ such that $f = 0$ on E_0 and $f = 1$ on E_1. ∎

The following theorem gives necessary and sufficient conditions for it to be possible to couple two random variables X and Y with values in E such that $X \leq Y$. A *coupling* of two random variables X and Y, in the most general sense of the word, is *a way to construct X and Y together on one underlying probability space* $(\Omega, \mathcal{F}, \mathbb{P})$. More precisely, if X and Y are random variables defined on different underlying probability spaces, then a coupling of X and Y is a pair of random variables (X', Y') defined on one underlying probability space $(\Omega, \mathcal{F}, \mathbb{P})$, such that X' is equally distributed with X and Y' is equally distributed with Y. Equivalently, since the laws of X and Y are all we really care about, we may say that a *coupling* of two probability laws μ, ν defined on measurable spaces $(E, \mathcal{E})$ and $(F, \mathcal{F})$, respectively, is a probability measure ρ on the product space $(E \times F, \mathcal{E} \otimes \mathcal{F})$ such that the first marginal of ρ is μ and its second marginal is ν. If two probability laws μ, ν satisfy the equivalent conditions of the following theorem, then we say that μ and ν are *stochastically ordered*, and we write[1] $\mu \leq \nu$.

Theorem 5.2 (Stochastic order) *Let E be a compact metrizable space that is equipped with a partial order that is compatible with the topology, and let μ, ν be probability laws on E. Then the following statements are equivalent:*

(i) $\int \mu(\mathrm{d}x) f(x) \leq \int \nu(\mathrm{d}x) f(x)\ \forall f \in C^+(E)$,
(ii) $\int \mu(\mathrm{d}x) f(x) \leq \int \nu(\mathrm{d}x) f(x)\ \forall f \in B^+(E)$,
(iii) *It is possible to couple random variables X, Y with laws $\mu = P[X \in \cdot\,]$ and $\nu = P[Y \in \cdot\,]$ in such a way that $X \leq Y$.*

Moreover, setting $\mu \leq \nu$ if and only if these conditions are satisfied defines a partial order on the space of probability measures on E.

Proof The implication (iii)⇒(ii) is easy: if X and Y are coupled such that $X \leq Y$ and f is monotone, then

$$\mathbb{E}\big[f(Y)\big] - \mathbb{E}\big[f(X)\big] = \mathbb{E}\big[f(Y) - f(X)\big] \geq 0,$$

since $f(Y) - f(X) \geq 0$ a.s. The implication (ii)⇒(i) is trivial. For the nontrivial implication (i)⇒(iii) we refer to [Lig85, Theorem II.2.4]. For finite spaces, a nice intuitive proof based on the max flow min cut theorem can be found in [Pre74]. To see that (i)–(iii) defines a partial order on the space of probability measures on E, we must check that 1. $\mu \leq \mu$, 2. $\mu \leq \nu$ and

[1] This notation may look a bit confusing at first sight, since, if μ, ν are probability measures on a measurable space $(\Omega, \mathcal{F})$, then one might interpret $\mu \leq \nu$ in a pointwise sense, that is, in the sense that $\mu(A) \leq \nu(A)$ for all $A \in \mathcal{F}$. In practice, this does not lead to confusion, since a pointwise inequality for probability measures is a very uninteresting property. Indeed, it is easy to check that probability measures μ, ν satisfy $\mu \leq \nu$ in a pointwise sense if and only if $\mu = \nu$.

$\nu \leq \mu$ imply $\mu = \nu$, and 3. $\mu \leq \nu \leq \rho$ implies $\mu \leq \rho$. Properties 1 and 3 are immediate from condition (i). Property 2 follows by combining (i) with Lemma 5.1. ∎

Sometimes it is more convenient (or intuitive) to work with events than with real functions. A set A is called *increasing* if its indicator function 1_A is monotone.

Lemma 5.3 (Increasing events) *Let E be a compact metrizable space that is equipped with a partial order that is compatible with the topology, and let μ, ν be probability measures on E. Then $\mu \leq \nu$ if and only if*

$$\mu(A) \leq \nu(A) \quad \textit{for all closed increasing } A \subset E. \tag{5.1}$$

Proof Condition (ii) of Theorem 5.2 clearly implies (5.1). To prove the converse, by condition (i) of Theorem 5.2, it suffices to show that (5.1) implies that $\int \mu(\mathrm{d}x)f(x) \leq \int \nu(\mathrm{d}x)f(x)$ for all $f \in \mathcal{C}^+(E)$. Fix $f \in \mathcal{C}^+(E)$. By adding a constant and multiplying with a positive constant we can without loss of generality assume that f takes values in $[0, 1]$. Define sets of dyadic rationals by $D_n := \{k2^{-n} : 0 \leq k \leq 2^n\}$ $(n \geq 1)$ and set $f_n(x) := \sup\{d \in D_n : d \leq f(x)\}$. Then

$$f_n = 2^{-n} \sum_{k=1}^{2^n} 1_{A_{n,k}} \quad \text{with} \quad A_{n,k} := \{x \in E : k2^{-n} \leq f(x)\}.$$

The sets $A_{n,k}$ are closed and increasing, so (5.1) implies that $\int \mu(\mathrm{d}x)f_n(x) \leq \int \nu(\mathrm{d}x)f_n(x)$ for all n. Since $f_n \uparrow f$, the claim follows. ∎

Exercise 5.4 *Let $\mathcal{C}^+(\mathbb{R}^d)$ denote the space of bounded continuous functions $f: \mathbb{R}^d \to \mathbb{R}$ that are monotone with respect to the product order on $\mathbb{R}^d$. It is well-known [KKO77] that for two probability measures μ, ν on $\mathbb{R}^d$ the following statements are equivalent: (i) $\int \mu(\mathrm{d}x)f(x) \leq \int \nu(\mathrm{d}x)f(x)$ for all $f \in \mathcal{C}^+(E)$, (ii) it is possible to couple random variables X, Y with laws $\mu = P[X \in \cdot\,]$ and $\nu = P[Y \in \cdot\,]$ in such a way that $X \leq Y$. Give a proof of this fact using Theorem 5.2.* Hint: *compactify.*

5.2 Monotone Interacting Particle Systems

In this section, we restrict ourselves to spaces of the form S^Λ, where S is a finite partially ordered set and Λ is countable. In particular, since Λ can be a set with only one element, this includes arbitrary finite partially ordered sets. We equip S^Λ with the product topology and product partial order. We start with a simple observation.

Lemma 5.5 (Compatibility of the product order) *Let S be a finite partially ordered set and let Λ be countable. Then the product partial order on S^Λ is compatible with the product topology on S^Λ.*

Proof Assume that $x_n, y_n, x, y \in S^\Lambda$ satisfy $x_n \leq y_n$ for all n and $x_n \to x$ and $y_n \to y$ in the product topology. Then, for each $i \in \Lambda$, there exists an N such that $x_n(i) = x(i)$ and $y_n(i) = y(i)$ for all $n \geq N$, and hence $x(i) \leq y(i)$. ∎

Because of Lemma 5.5, Theorem 5.2 is applicable with $E = S^\Lambda$. The following lemma shows that for E of this form, in condition (i) of Theorem 5.2, we can replace $\mathcal{C}^+(S^\Lambda)$ by

$C^+_{\mathrm{fin}}(S^\Lambda) := C^+(S^\Lambda) \cap C_{\mathrm{fin}}(S^\Lambda)$, the space of monotone functions $f\colon S^\Lambda \to \mathbb{R}$ that depend on finitely many coordinates.

Lemma 5.6 (Local monotone functions) *The space $C^+_{\mathrm{fin}}(S^\Lambda)$ is dense in $C^+(S^\Lambda)$.*

Proof Fix $z \in S^\Lambda$, choose finite $\Lambda_n \uparrow \Lambda$, and for each $z \in S^\Lambda$ define $x^n(i) := x(i)$ if $i \in \Lambda_n$ and $:= z(i)$ otherwise. Fix $f \in C^+(S^\Lambda)$ and define $f_n(x) := f(x^n)$. Then clearly $f_n \in C^+_{\mathrm{fin}}(S^\Lambda)$ for all n. The argument used at the beginning of the proof of Lemma 4.31 shows that $\|f_n - f\|_\infty \to 0$, completing the proof. ∎

By definition, a probability kernel K on S^Λ is *monotone* if it satisfies the following equivalent conditions. Note that in (i), $\leq$ denotes the stochastic order. The equivalence of (i)–(iii) is a trivial consequence of Theorem 5.2.

(i) $K(x, \cdot\,) \leq K(y, \cdot\,)$ for all $x \leq y$.
(ii) $Kf \in B^+(S^\Lambda)$ for all $f \in C^+(S^\Lambda)$.
(iii) $Kf \in B^+(S^\Lambda)$ for all $f \in B^+(S^\Lambda)$.

We note that if K is monotone, then

$$\mu \leq \nu \quad \text{implies} \quad \mu K \leq \nu K. \tag{5.2}$$

Indeed, this follows from (iii) since $f \in B^+(S^\Lambda)$ implies $Kf \in B^+(S^\Lambda)$ and hence $\mu K f \leq \nu K f$ since $\mu \leq \nu$.

Recall from (2.3) that a *random mapping representation* of a probability kernel K is a random map M such that

$$K(x, \cdot\,) = \mathbb{P}[M(x) \in \cdot\,] \quad \forall x. \tag{5.3}$$

We say that K can be represented in the class of monotone maps, or that K is *monotonically representable*, if there exists a random monotone map M such that (5.3) holds. In Chapter 4, we based our construction of an interacting particle system on a random mapping representation of its generator G in terms of continuous maps, of the form

$$Gf(x) = \sum_{m \in \mathcal{G}} r_m \big\{ f\big(m(x)\big) - f\big(x\big) \big\}, \tag{5.4}$$

where the rates satisfy (4.11) or possibly the weaker conditions from Theorem 4.18. If there exists such a random mapping representation for which all maps $m \in \mathcal{G}$ are monotone, then we say that G is *monotonically representable*.

Lemma 5.7 (Monotone representability) *Each monotonically representable probability kernel is monotone. If the generator of an interacting particle system is monotonically representable, then, for each $t \geq 0$, the transition probability P_t is a monotonically representable probability kernel.*

Proof If a probability kernel K can be written in the form (5.3) with M a random monotone map, then for each $x \leq y$, the random variables $M(x)$ and $M(y)$ are coupled such that $M(x) \leq M(y)$ a.s., so their laws are stochastically ordered as $K(x, \cdot\,) \leq K(y, \cdot\,)$. Since this holds for all $x \leq y$, the kernel K is monotone.

Given a random mapping representation of the form (5.4) of the generator G of an interacting particle system, we can construct a stochastic flow $(\mathbf{X}_{s,t})_{s \leq t}$ as in Theorem 4.14

based on a graphical representation ω. If all maps $m \in \mathcal{G}$ are monotone, then for each finite $\omega' \subset \omega$, the maps $(\mathbf{X}^{\omega'}_{s,t})_{s\le t}$ defined in (4.27) are also monotone, since they are the concatenation of finitely many maps from $\mathcal{G}$. By Proposition 4.23, this implies that the maps $(\mathbf{X}_{s,t})_{s\le t}$ are also monotone. It follows that

$$P_t(x,\,\cdot\,) = \mathbb{P}\big[\mathbf{X}_{0,t}(x) \in \cdot\,\big]$$

is a representation of P_t in terms of the random monotone map $\mathbf{X}_{0,t}$, so P_t is monotonically representable. ∎

We say that an interacting particle system is *monotone* if its transition kernels are monotone probability kernels, and we say that it is *monotonically representable* if its generator is monotonically representable. Somewhat surprisingly, it turns out that for probability kernels, "monotonically representable" is a strictly stronger concept than being "monotone". See [FM01] for an example of a probability kernel on $\{0,1\}^2$ that is monotone but not monotonically representable. Nevertheless, it turns out that (almost) all monotone interacting particle systems that one encounters in practice are also monotonically representable.

The following maps are examples of monotone maps:

- The voter map $\texttt{vot}_{ij}$ defined in (1.4).
- The branching map $\texttt{bra}_{ij}$ defined in (1.6).
- The death map $\texttt{death}_i$ defined in (1.7).
- The asymmetric exclusion map $\texttt{asep}_{ij}$ defined in (1.20).
- The exclusion map $\texttt{excl}_{ij}$ defined in (1.22).
- The coalescing random walk map $\texttt{rw}_{ij}$ defined in (1.24).
- The cooperative branching map $\texttt{coop}_{ij}$ defined in (1.27).
- The maps $m_{i,L}$ defined in (4.47) to construct the Ising model with Glauber dynamics.

As a result, the following interacting particle systems are monotonically representable (and hence, in particular, monotone):

- The voter model with generator as in (1.5).
- The contact process with generator as in (1.8).
- The ferromagnetic Ising model with Glauber dynamics, since its generator can be written as in (4.48).
- The biased voter model with generator as in (1.17).
- The exclusion process with generator as in (1.23).
- Systems of coalescing random walks with generator as in (1.25).
- Systems with cooperative branching and coalescence as in Figure 1.11.

On the other hand, the following maps are *not* monotone:

- The annihilating random walk map $\texttt{arw}_{ij}$ defined in (1.26).
- The killing map $\texttt{kill}_{ij}$ defined in (1.28).

Examples of interacting particle systems that are not monotone[2] are:

[2] Note that the fact that a given interacting particle system is represented in maps that are not monotone does not prove that the system is not monotone. Indeed, it is conceivable that the same system can also be monotonically represented. See Exercises 6.17 and 6.18 for an interacting particle system that has a monotone and a non-monotone random mapping representation, both of which are useful.

- The antiferromagnetic Ising model with Glauber dynamics.
- The Neuhauser–Pacala model from (1.18) for small values of α.
- Systems of annihilating random walks.
- The biased annihilating branching process of (1.29).

5.3 Positive Correlations

In this section, we study positive correlations. Positive correlations play an important role in many more involved arguments, but unfortunately we will not see an example of this, so readers may skip the present section at an initial reading. Let S be a finite partially ordered set and let Λ be countable. A probability measure μ on S^Λ has *positive correlations* if it satisfies the equivalent conditions of the following lemma. Recall from (2.9) that $\mathrm{Cov}_\mu(f,g)$ denotes the covariance of f and g under μ.

Lemma 5.8 (Positive correlations) *Let S be a finite partially ordered set, let Λ be countable, and let μ be a probability measure on S^Λ. Then the following conditions are equivalent:*

(i) $\mathrm{Cov}_\mu(f,g) \geq 0 \quad \forall f,g \in C^+_{\mathrm{fin}}(S^\Lambda)$,
(ii) $\mathrm{Cov}_\mu(f,g) \geq 0 \quad \forall f,g \in B^+(S^\Lambda)$.

Proof We introduce the following notation. For any $f \geq 0$ with $\mu f > 0$, we let μ_f denote the probability measure on S^Λ defined as $\mu_f(g) := \mu(fg)/\mu f$ $(g \in B(S^\Lambda))$. In particular, if f is the indicator function of an event, then μ_f is the law obtained from μ by conditioning on this event. If $f,g \in B(S^\Lambda)$, $f \geq 0$, and $\mu f > 0$, then

$$\mathrm{Cov}_\mu(f,g) \geq 0 \quad \Leftrightarrow \quad \mu(fg) \geq (\mu f)(\mu g) \quad \Leftrightarrow \quad \mu_f g \geq \mu g.$$

We now prove the equivalence of (i) and (ii). Trivially, (ii) implies (i). Conversely, if (i) holds, then for each $f,g \in C^+_{\mathrm{fin}}(S^\Lambda)$ with $f \geq 1$, one has $\mu_f g \geq \mu g$, which using Lemma 5.6 implies that $\mu_f \geq \mu$ in the stochastic order. By Theorem 5.2, this implies that $\mu_f g \geq \mu g$ for all $g \in B^+(S^\Lambda)$ and hence $\mathrm{Cov}_\mu(f,g) \geq 0$ for all $f \in C^+_{\mathrm{fin}}(S^\Lambda)$ with $f \geq 1$ and $g \in B^+(S^\Lambda)$. Since adding a constant to f does not change the covariance, we can remove the condition $f \geq 1$. Repeating the argument with the roles of f and g interchanged then yields (ii). ∎

Exercise 5.9 *In the context of Lemma 5.8, show that a probability measure μ on S^Λ has positive correlations if and only if $\mu(A|B) \geq \mu(A)$ for all measurable increasing events $A,B \subset S^\Lambda$ such that $\mu(B) > 0$. Show that it suffices to check this condition for events that depend on finitely many coordinates.* Hint: *Lemma 5.3.*

Exercise 5.10 *Show that a probability measure μ on $\{0,1\}^2$ has positive correlations if and only if $\mu(00)\mu(11) \geq \mu(01)\mu(10)$. For probability measures on $\{0,1\}^n$ with $n \geq 3$, it is known that the FKG condition $\mu(x \wedge y)\mu(x \vee y) \geq \mu(x)\mu(y)$ implies (but is not equivalent to) positive correlations, see [FKG71].*

The following proposition gives sufficient conditions for the time evolution of an interacting particle system to preserve the space of probability measures with positive correlations. Condition (i) says that the interacting particle system is monotone. Note that we do not assume monotone representability. Condition (ii) says that the system only jumps between comparable states. Note that this condition is actually independent of the graphical representation (assuming all rates r_m are strictly positive). Variations of this result can be found in [Har77, Cox84] and [Lig85, Thm II.2.14].

Proposition 5.11 (Preservation of positive correlations) *Let S be a finite partially ordered set, let Λ be countable, and let $(P_t)_{t\geq 0}$ be the semigroup of an interacting particle system with generator of the form (4.1) with the rates satisfying (4.11). Assume that*

(i) *P_t is monotone for each $t \geq 0$,*
(ii) *for each $x \in S^\Lambda$ and $m \in \mathcal{G}$, either $x \leq m(x)$ or $x \geq m(x)$.*

Assume that μ is a probability measure on S^Λ with positive correlations. Then μP_t has positive correlations for each $t \geq 0$.

Proof Let $C^+_{\rm sum} = C^+_{\rm sum}(S^\Lambda) := C^+(S^\Lambda) \cap C_{\rm sum}(S^\Lambda)$. By Lemma 4.33, P_t maps $C_{\rm sum}$ into itself for each $t \geq 0$. Condition (i) then implies that P_t maps $C^+_{\rm sum}$ into itself. Condition (ii) and formula (4.39) imply that

$$\Gamma_{\mathcal{G}}(f,g) = \sum_{m\in\mathcal{G}} r_m\{f(m(x)) - f(x)\}\{g(m(x)) - g(x)\} \geq 0$$

for all $f, g \in C^+_{\rm sum}$. Proposition 4.35 tells us that for $f, g \in C^+_{\rm sum}$,

$$\mathrm{Cov}_{\mu P_t}(f,g) = \mathrm{Cov}_\mu(P_t f, P_t g) + \int_0^t \mathrm{d}s\, \mu P_{t-s}\Gamma_{\mathcal{G}}(P_s f, P_s g).$$

Using the fact that μ has positive correlations and our previous observations, we see that the right-hand side of this equation is nonnegative, proving that μP_t has positive correlations. ∎

Exercise 5.12 *Let Λ be a countable set and let $n \geq 1$. Show that product measures on $\{0,\dots,n\}^\Lambda$ have positive correlations.* Hint: *construct an interacting particle system that has the desired product measure as its invariant law.*

5.4 The Upper and Lower Invariant Laws

In the present section, we assume that the local state space is $S = \{0, 1\}$, which covers all examples of monotone interacting particle systems mentioned in Section 5.2. We will use the phrase "an interacting particle system with state space $\{0,1\}^\Lambda$" as a shorthand for any interacting particle that can be constructed from a graphical representation with rates that satisfy (4.11) or possibly the weaker conditions from Theorem 4.18. We use the symbols $\underline{0}$ and $\underline{1}$ to denote the states in S^Λ that are identically 0 or 1, respectively. Below, $\delta_{\underline{0}}$ denotes the delta measure at the configuration that is identically 0, so $\delta_{\underline{0}} P_t$ denotes the law at time t of the process started in $X_0(i) = 0$ a.s. $(i \in \Lambda)$.

Theorem 5.13 (Upper and lower invariant laws) *Let X be an interacting particle system with state space of the form $\{0,1\}^\Lambda$ and transition probabilities $(P_t)_{t\geq 0}$. Assume that X is monotone. Then there exist invariant laws $\underline{\nu}$ and $\overline{\nu}$ such that*

$$\delta_{\underline{0}} P_t \underset{t\to\infty}{\Longrightarrow} \underline{\nu} \quad \text{and} \quad \delta_{\underline{1}} P_t \underset{t\to\infty}{\Longrightarrow} \overline{\nu}.$$

If ν is any other invariant law, then $\underline{\nu} \leq \nu \leq \overline{\nu}$.

The invariant laws $\underline{\nu}$ and $\overline{\nu}$ from Theorem 5.13 are called *lower* and *upper invariant laws*, respectively. Before we give the proof of Theorem 5.13, we start with two preparatory lemmas.

Lemma 5.14 (Equal mean) *Let μ, ν be probability laws on $\{0,1\}^\Lambda$ such that $\mu \leq \nu$ and*

$$\int \mu(\mathrm{d}x)\,x(i) \geq \int \nu(\mathrm{d}x)\,x(i) \qquad (i \in \Lambda).$$

Then $\mu = \nu$.

Proof By Theorem 5.2, we can couple random variables with laws $\mathbb{P}[X \in \cdot\,] = \mu$ and $\mathbb{P}[Y \in \cdot\,] = \nu$ in such a way that $X \leq Y$. Now $\mathbb{E}[X(i)] \geq \mathbb{E}[Y(i)]$ implies $\mathbb{E}[Y(i) - X(i)] \leq 0$. Since $Y(i) - X(i) \geq 0$ a.s., it follows that $X(i) = Y(i)$ a.s. In particular, if this holds for all $i \in \Lambda$, then $\mu = \nu$. ∎

Lemma 5.15 (Monotone convergence of probability laws) *Let $(\nu_n)_{n\geq 0}$ be a sequence of probability laws on $\{0,1\}^\Lambda$ that are stochastically ordered as $\nu_k \leq \nu_{k+1}$ $(k \geq 0)$. Then there exists a probability law ν on $\{0,1\}^\Lambda$ such that $\nu_n \Rightarrow \nu$, that is, the ν_n converge weakly to ν.*

Proof Since $\nu_n f$ increases to a finite limit for each $f \in \mathcal{C}^+(\{0,1\}^\Lambda)$, this is an immediate consequence of Lemmas 5.1 and 4.40. ∎

Proof of Theorem 5.13 By symmetry, it suffices to prove the statement for $\underline{\nu}$. Since $\underline{0}$ is the lowest possible state, for each $t \geq 0$, we trivially have

$$\delta_{\underline{0}} \leq \delta_{\underline{0}} P_t$$

By (5.2), this implies that

$$\delta_{\underline{0}} P_s \leq \delta_{\underline{0}} P_t P_s = \delta_{\underline{0}} P_{t+s} \qquad (s, t \geq 0),$$

which shows that $t \mapsto \delta_{\underline{0}} P_t$ is nondecreasing with respect to the stochastic order. By Lemma 5.15, each monotone sequence of probability laws has a weak limit, so there exists a probability law $\underline{\nu}$ on $\{0,1\}^\Lambda$ such that

$$\delta_{\underline{0}} P_t \underset{t\to\infty}{\Longrightarrow} \underline{\nu}.$$

It follows from Lemma 4.42 that $\underline{\nu}$ is an invariant law.

To complete the proof of the theorem, we observe that if ν is any other invariant law, then, by (5.2),

$$\delta_{\underline{0}} \leq \nu \quad \Rightarrow \quad \delta_{\underline{0}} P_t \leq \nu P_t = \nu \quad (t \geq 0).$$

Since $\delta_{\underline{0}} P_t \Rightarrow \nu$ as $t \to \infty$, if follows that $\underline{\nu} f \leq \nu f$ for all $f \in \mathcal{C}^+(\{0,1\}^\Lambda)$, which by Theorem 5.2 implies that $\underline{\nu} \leq \nu$. ∎

Theorem 5.16 (Ergodicity of monotone systems) *Let X be a monotone interacting particle system with state space $\{0,1\}^\Lambda$ and lower and upper invariant laws $\underline{\nu}$ and $\overline{\nu}$. If*

$$\int \underline{\nu}(\mathrm{d}x)x(i) = \int \overline{\nu}(\mathrm{d}x)x(i) \qquad \forall i \in \Lambda, \tag{5.5}$$

then X has a unique invariant law $\nu := \underline{\nu} = \overline{\nu}$ and is ergodic in the sense that

$$\mathbb{P}^x\big[X_t \in \cdot\,\big] \underset{t\to\infty}{\Longrightarrow} \nu \qquad (x \in \{0,1\}^\Lambda).$$

On the other hand, if (5.5) does not hold, then X has at least two invariant laws.

Proof By Lemma 5.14, (5.5) is equivalent to the condition that $\underline{\nu} = \overline{\nu}$. It is clear that if $\underline{\nu} \neq \overline{\nu}$, then X has at least two invariant laws and ergodicity cannot hold. On the other hand, by Theorem 5.13, any invariant law ν satisfies $\underline{\nu} \leq \nu \leq \overline{\nu}$, so if $\underline{\nu} = \overline{\nu}$, then $\nu = \underline{\nu} = \overline{\nu}$.

To complete the proof, we must show that $\underline{\nu} = \overline{\nu} =: \nu$ implies $\delta_x P_t \Rightarrow \nu$ as $t \to \infty$ for all $x \in \{0,1\}^\Lambda$. Since

$$\delta_{\underline{0}} P_t f \leq \delta_x P_t f \leq \delta_{\underline{1}} P_t f$$

for all $f \in C^+(\{0,1\}^\Lambda)$, we see that

$$\underline{\nu} f \leq \liminf_{t\to\infty} P_t f \leq \limsup_{t\to\infty} P_t f \leq \overline{\nu} f$$

for all $f \in C^+(\{0,1\}^\Lambda)$. The claim now follows from Lemmas 4.40 and 5.1. ∎

To state the final result of this section, we need a bit of theory. We observe that for any interacting particle system, the set $\mathcal{I}$ of all invariant laws is a compact, convex subset of the space $\mathcal{M}_1(S^\Lambda)$ of probability measures on S^Λ, equipped with the topology of weak convergence. Indeed, if μ and ν are invariant laws and $p \in [0,1]$, then clearly

$$\big(p\mu + (1-p)\nu\big)P_t = p\mu P_t + (1-p)\nu P_t = p\mu + (1-p)\nu \quad (t \geq 0),$$

proving that $p\mu + (1-p)\nu$ is an invariant law. The fact that $\mathcal{I}$ is closed follows from Proposition 4.45. Since $\mathcal{M}_1(S^\Lambda)$ is compact, $\mathcal{I}$ is also compact.

By definition, an element $\nu \in \mathcal{I}$ is called *extremal* if it cannot be written as a nontrivial convex combination of other elements of $\mathcal{I}$, that is,

$$\nu = p\nu_1 + (1-p)\nu_2 \qquad (0 < p < 1,\ \nu_1, \nu_2 \in \mathcal{I}) \quad \text{implies} \quad \nu_1 = \nu_2 = \nu.$$

We let

$$\mathcal{I}_{\mathrm{e}} := \{\nu \in \mathcal{I} : \nu \text{ is an extremal element of } \mathcal{I}\}.$$

Since $\mathcal{I}$ is compact and convex, Choquet's theorem implies that each invariant law ν can be written as

$$\nu = \int_{\mathcal{I}_{\mathrm{e}}} \rho_\nu(\mathrm{d}\mu)\mu,$$

where ρ_ν is a probability measure on $\mathcal{I}_{\mathrm{e}}$. In practice, it happens quite often[3] that $\mathcal{I}_{\mathrm{e}}$ is a finite set.[4] In this case, Choquet's theorem simply says that each invariant law is a convex combination of the extremal invariant laws; that is, each invariant law is of the form

$$\nu = \sum_{\mu \in \mathcal{I}_{\mathrm{e}}} p(\mu)\mu,$$

[3] The the voter model in dimensions $d \geq 3$ is a counterexample. The Ising model in dimensions $d \geq 3$ is also a counterexample, although for the Ising model, it is still true that $\underline{\nu}$ and $\overline{\nu}$ are the only extremal invariant measures that are moreover translation invariant.

[4] This may, however, be quite difficult to prove!

where $(p(\mu))_{\mu\in\mathcal{I}_e}$ are nonnegative constants, summing up to one. In view of this, we are naturally interested in finding all extremal invariant laws of a given interacting particle system.

Lemma 5.17 (The lower and upper invariant laws are extremal.) *Let X be a monotone interacting particle system with state space $\{0,1\}^\Lambda$ and lower and upper invariant laws $\underline{\nu}$ and $\overline{\nu}$. Then $\underline{\nu}$ and $\overline{\nu}$ are extremal invariant laws of X.*

Proof By symmetry, it suffices to prove the statement for $\overline{\nu}$. Imagine that

$$\overline{\nu} = p\nu_1 + (1-p)\nu_2 \quad \text{for some} \quad 0 < p < 1,\ \nu_1, \nu_2 \in \mathcal{I}.$$

By Theorem 5.13, for each $f \in B^+(\{0,1\}^\Lambda)$, one has $\nu_1 f \leq \overline{\nu} f$ and $\nu_2 f \leq \overline{\nu} f$. Since

$$p(\overline{\nu} f - \nu_1 f) + (1-p)(\overline{\nu} f - \nu_2 f) = 0,$$

it follows that $\overline{\nu} f = \nu_1 f = \nu_2 f$. Since this holds for each monotone f, we conclude (by Lemma 5.1) that $\overline{\nu} = \nu_1 = \nu_2$. ∎

Exercise 5.18 *Let X be an interacting particle system with state space $\{0,1\}^\Lambda$ and generator G. Assume that G has a random mapping representation in terms of monotone maps and let $(\mathbf{X}_{s,t})_{s\leq t}$ be the corresponding stochastic flow as in Theorem 4.14. Show that the a.s. limits*

$$\left.\begin{aligned} \underline{X}_t &:= \lim_{s\to-\infty} \mathbf{X}_{s,t}(\underline{0}), \\ \overline{X}_t &:= \lim_{s\to-\infty} \mathbf{X}_{s,t}(\underline{1}) \end{aligned}\right\} \quad (t \in \mathbb{R})$$

define stationary Markov processes $(\underline{X}_t)_{t\in\mathbb{R}}$ and $(\overline{X}_t)_{t\in\mathbb{R}}$ whose invariant laws

$$\underline{\nu} = \mathbb{P}[\underline{X}_t \in \cdot\,] \quad \textit{and} \quad \overline{\nu} = \mathbb{P}[\overline{X}_t \in \cdot\,] \qquad (t \in \mathbb{R})$$

are the lower and upper invariant laws of X, respectively. Show that (5.5) implies that

$$\lim_{s\to-\infty} \mathbf{X}_{s,t}(x) = \underline{X}_t = \overline{X}_t \quad \text{a.s.} \quad (x \in \{0,1\}^\Lambda,\ t \in \mathbb{R}).$$

5.5 The Contact Process

The contact process has been defined on $\mathbb{Z}^d$ in (1.8) and on a very general class of lattices in (2.45). In the present section, we will look at a class of contact processes that are more general than those in (1.8) but a bit less general than those in (2.45). Throughout this section, Λ will be a countable set, and p will be a probability kernel on Λ that is *symmetric* in the sense that $p(i,j) = p(j,i)$ $(i,j \in \Lambda)$ and satisfies $p(i,i) = 0$ $(i \in \Lambda)$. By definition, an *automorphism* of (Λ, p) is a bijection $\psi : \Lambda \to \Lambda$ such that $p(\psi(i), \psi(j)) = p(i,j)$ $(i,j \in \Lambda)$ (compare footnote 8 on page 50). We will assume that (Λ, p) is *vertex transitive* in the sense that

$$\forall i,j \in \Lambda\ \exists \text{ automorphism } \psi \text{ of } (\Lambda, p) \text{ s.t. } \psi(i) = j. \tag{5.6}$$

We will be interested in contact processes with a generator of the form

$$G_{\text{cont}} f(x) := \lambda \sum_{i,j\in\Lambda} p(i,j)\big\{f\big(\texttt{bra}_{ij}(x)\big) - f\big(x\big)\big\} + \sum_{i\in\Lambda} \big\{f\big(\texttt{death}_i(x)\big) - f\big(x\big)\big\}, \tag{5.7}$$

where $\lambda \geq 0$ is the *infection rate* and the death rate is one. Note that our present definition differs a bit from the classical definition of the contact process on $\mathbb{Z}^d$ in (1.8) in the sense that in (5.7) the total rate of all infections out of a site i is λ, while in (1.8) it is $\lambda|\mathcal{N}_i|$, where $|\mathcal{N}_i|$ is the number of neighbors of i. Already when we studied the mean-field limit of the contact process, we discovered that the normalization in (5.7) is often more natural, compare (3.21).

Since both the branching and death map are monotone, the contact process is a monotonically representable interacting particle system, so by Theorem 5.13, it has a lower and upper invariant law $\underline{\nu}$ and $\overline{\nu}$. Since $\mathtt{bra}_{ij}(\underline{0}) = \underline{0}$ and $\mathtt{death}_i(\underline{0}) = \underline{0}$ for each $i, j \in \Lambda$, the all-zero configuration $\underline{0}$ is a trap for the contact process, so $\delta_{\underline{0}} P_t = \delta_{\underline{0}}$ for all $t \geq 0$, and hence

$$\underline{\nu} = \delta_{\underline{0}}.$$

Therefore, by Theorem 5.16, the contact process is ergodic if and only if the function

$$\theta(\lambda) := \int \overline{\nu}_\lambda(\mathrm{d}x)\, x(i) \qquad (i \in \mathbb{Z}^d) \tag{5.8}$$

satisfies $\theta(\lambda) = 0$. Here $\overline{\nu}_\lambda$ denotes the upper invariant law of the contact process with infection rate λ and the right-hand side of (5.8) does not depend on $i \in \Lambda$ by our assumption that (Λ, p) is vertex transitive. For reasons that will become clear in the next chapter (see Lemma 6.29), $\theta(\lambda)$ is actually the same as the survival probability started from a single occupied site; that is, this is the function in Figure 1.4.

By definition, we say that a probability law μ on $\{0,1\}^\Lambda$ is *nontrivial* if

$$\mu(\{\underline{0}\}) = 0,$$

that is, if μ gives zero probability to the all-zero configuration.

Lemma 5.19 (Nontriviality of the upper invariant law) *For the contact process, if $\overline{\nu} \neq \delta_{\underline{0}}$, then $\overline{\nu}$ is nontrivial.*

Proof We can always write $\overline{\nu} = (1-p)\delta_{\underline{0}} + p\mu$ where $p \in [0,1]$ and μ is a nontrivial law. By assumption, $\overline{\nu} \neq \delta_{\underline{0}}$, so $p > 0$. Since $\overline{\nu}$ and $\delta_{\underline{0}}$ are invariant laws, μ must be an invariant law too. By Lemma 5.17, $\overline{\nu}$ cannot be written as a nontrivial convex combination of other invariant laws, so we conclude that $p = 1$. ∎

Proposition 5.20 (Monotonicity in the infection rate) *Let $\overline{\nu}_\lambda$ denote the upper invariant law of the contact process with infection rate λ. Then $\lambda \leq \lambda'$ implies $\overline{\nu}_\lambda \leq \overline{\nu}_{\lambda'}$. In particular, the function $\lambda \mapsto \theta(\lambda)$ is nondecreasing.*

Proof Let X and X' be contact processes started in the initial state $X_0 = \underline{1} = X'_0$ and with infection rates λ and λ'. It suffices to prove that X and X' can be coupled such that $X_t \leq X'_t$ for all $t \geq 0$.

We will couple the graphical representations of the processes with infection rates λ and λ'. We write $\mathcal{G} = \mathcal{G}_{\mathrm{bra}} \cup \mathcal{G}_{\mathrm{death}}$, where

$$\mathcal{G}_{\mathrm{bra}} := \{\mathtt{bra}_{ij} : i, j \in \Lambda\} \quad \text{and} \quad \mathcal{G}_{\mathrm{death}} := \{\mathtt{death}_i : i \in \Lambda\}.$$

Then X can be constructed as in Theorem 4.14 from a Poisson point set ω on $\mathcal{G} \times \mathbb{R}$ with intensity measure ρ_λ given by

$$\left.\begin{aligned}\rho_\lambda(\{\mathtt{bra}_{ij}\}\times[s,t]) &:= \lambda p(i,j)(t-s),\\ \rho_\lambda(\{\mathtt{death}_i\}\times[s,t]) &:= (t-s),\end{aligned}\right\}\quad (i,j\in\Lambda,\ s\le t).$$

Likewise, X' can be constructed from a Poisson point set ω' with intensity $\rho_{\lambda'}$. We claim that we can couple ω and ω' in such a way that the latter has more branching maps and the same death maps as ω. This can be done as follows. Let ω'' be a Poisson point set on $\mathcal{G}\times\mathbb{R}$, independent of ω, with intensity measure $\rho'' := \rho_{\lambda'} - \rho_\lambda$, that is,

$$\left.\begin{aligned}\rho''(\{\mathtt{bra}_{ij}\}\times[s,t]) &:= (\lambda'-\lambda)p(i,j)(t-s),\\ \rho''(\{\mathtt{death}_i\}\times[s,t]) &:= 0,\end{aligned}\right\}\quad (i,j\in\Lambda,\ s\le t).$$

Since the sum of two independent Poisson sets yields another Poisson set, setting

$$\omega' := \omega + \omega''$$

defines a Poisson point set with intensity $\rho_{\lambda'}$. We observe that

$$\begin{aligned} x\le x' &\quad\text{implies}\quad \mathtt{bra}_{ij}(x)\le\mathtt{bra}_{ij}(x'),\\ x\le x' &\quad\text{implies}\quad \mathtt{death}_i(x)\le\mathtt{death}_i(x'),\\ x\le x' &\quad\text{implies}\quad x\le\mathtt{bra}_{ij}(x'). \end{aligned}$$

The first two statements just say that the maps $\mathtt{bra}_{ij}$ and $\mathtt{death}_i$ are monotone. The third statement says that if we apply a branching map only to the larger configuration x', then the order between x and x' is preserved.

Since ω' has the same branching and death maps as ω, plus some extra branching maps, using Proposition 4.23, we conclude that the stochastic flows $(\mathbf{X}_{s,t})_{s\le t}$ and $(\mathbf{X}'_{s,t})_{s\le t}$ constructed from ω and ω' satisfy

$$x\le x' \quad\text{implies}\quad \mathbf{X}_{s,t}(x)\le\mathbf{X}'_{s,t}(x')\qquad (s\le t).$$

In particular, setting $X_t := \mathbf{X}_{0,t}(\underline{1})$ and $X'_t := \mathbf{X}'_{0,t}(\underline{1})$ yields the desired coupling between X and X'. ∎

Exercise 5.21 *For contact processes with generator of the form (5.7), calculate the constant $K_\downarrow$ from (4.16) and apply Theorem 4.36 to conclude that*

$$\lambda < 1 \quad\textit{implies}\quad \overline{\nu} = \delta_{\underline{0}}.$$

In Chapter 7, we will prove that $\theta(\lambda) > 0$ for λ sufficiently large.

5.6 Other Examples

The Ising Model with Glauber Dynamics

We have seen in (4.48) that the generator of the Ising model with Glauber dynamics is monotonically representable, so by Theorem 5.13,[5] it has a lower and upper invariant law $\underline{\nu}$

[5] The difference between the local state space $\{-1, 1\}$ of the Ising model and $\{0, 1\}$ of Theorem 5.13 is, of course, entirely notational.

and $\overline{\nu}$. We let

$$m_*(\beta) := \int \overline{\nu}(\mathrm{d}x)\, x(i),$$

which is independent of i if the process has some translation invariant structure (like the nearest neighbor or range R processes on $\mathbb{Z}^d$). For reasons that cannot be explained here, this function is actually the same as the one defined in (1.15); that is, this is the *spontaneous magnetization* of the Ising model; see Figure 1.6. By the symmetry between -1 and $+1$ spins, we clearly have

$$\int \underline{\nu}(\mathrm{d}x)\, x(i) = -m_*(\beta).$$

By Theorem 4.37, we have

$$\beta < N^{-1}\big(\log(N+1) - \log(N-1)\big) \quad \text{implies} \quad \underline{\nu} = \overline{\nu},$$

from which we conclude that $m_*(\beta) = 0$ for β sufficiently small,

The function $\beta \mapsto m_*(\beta)$ is nondecreasing, but this cannot be proved with the sort of techniques used in Proposition 5.20. The lower and upper invariant laws of the Ising model with Glauber dynamics are infinite volume Gibbs measures, and much of the analysis of the Ising model is based on this fact. In fact, the Ising model with Glauber dynamics is just one example of an interacting particle system that has these Gibbs measures as its invariant laws. In general, interacting particle systems with this property are called stochastic Ising models, and the Gibbs measures themselves are simply called the Ising model. We refer to [Lig85, Chapter IV] for an exposition of this material. In particular, in [Lig85, Thm IV.3.14], it is shown that for the nearest-neighbor model on $\mathbb{Z}^2$, one has $m_*(\beta) > 0$ for β sufficiently large.

The Voter Model

Consider a voter model with local state space $S = \{0, 1\}$. Since the voter maps $\mathtt{vot}_{ij}$ from (1.4) are monotone, the voter model is monotonically representable. Since both the constant configurations $\underline{0}$ and $\underline{1}$ are traps,

$$\underline{\nu} = \delta_{\underline{0}} \quad \text{and} \quad \overline{\nu} = \delta_{\underline{1}},$$

so we conclude (recall Theorem 5.16) that the voter model is never ergodic. For the model on $\mathbb{Z}^d$, it is proved in [Lig85, Thm V.1.8] that if $d = 1, 2$, then $\delta_{\underline{0}}$ and $\delta_{\underline{1}}$ are the only extremal invariant laws. On the other hand, in dimensions $d \geq 3$, the set $\mathcal{I}_{\mathrm{e}}$ of extremal invariant laws is of the form $\{\nu_p : p \in [0, 1]\}$, where the invariant measure ν_p has intensity $\int \nu_p(\mathrm{d}x)\, x(i) = p$. We will give a partial proof of these statements in Section 6.7.

5.7 Exercises

Exercise 5.22 *Give an example of two probability measures μ, ν on a set of the form $\{0, 1\}^\Lambda$ that satisfy*

$$\int \mu(\mathrm{d}x)x(i) \leq \int \nu(\mathrm{d}x)x(i) \qquad (i \in \Lambda),$$

but that are not *stochastically ordered as $\mu \leq \nu$.*

Exercise 5.23 *Let $(X_t^\lambda)_{t\geq 0}$ denote the contact process with infection rate λ (and death rate one), started in $X_0^\lambda = \underline{1}$. Apply Corollary 4.44 to prove that for each fixed $t \geq 0$, the function*

$$\theta_t(\lambda) := \mathbb{P}[\mathbf{X}_{0,t}^\lambda(\underline{1})(i) = 1] \tag{5.9}$$

depends continuously on λ. Use this to conclude that the function $\theta(\lambda)$ from (5.8) is right-continuous. Hint: *Use that the decreasing limit of continuous functions is upper semi-continuous.*

For the next exercise, let us define a *double death* map

$$\mathtt{death}_{ij}(x)(k) := \begin{cases} 0 & \text{if } k \in \{i, j\}, \\ x(k) & \text{otherwise.} \end{cases} \tag{5.10}$$

Recall the branching map $\mathtt{bra}_{ij}$ defined in (1.6), the death map $\mathtt{death}_i$ defined in (1.7), and the cooperative branching map $\mathtt{coop}_{ij}$ defined in (1.27). Consider the cooperative branching process X with values in $\{0, 1\}^{\mathbb{Z}}$ with generator

$$G_X f(x) = \lambda \sum_{i\in\mathbb{Z}} \sum_{\sigma\in\{-1,+1\}} \big\{f\big(\mathtt{coop}_{i+2\sigma,i+\sigma,i}(x)\big) - f\big(x\big)\big\} + \sum_{i\in\mathbb{Z}} \big\{f\big(\mathtt{death}_i(x)\big) - f\big(x\big)\big\},$$

and the contact process with double deaths Y with generator

$$G_Y f(y) = \lambda \sum_{i\in\mathbb{Z}} \sum_{\sigma\in\{-1,+1\}} \big\{f\big(\mathtt{bra}_{i+\sigma,i}(y)\big) - f\big(y\big)\big\} + \sum_{i\in\mathbb{Z}} \big\{f\big(\mathtt{death}_{i,i+1}(y)\big) - f\big(y\big)\big\}.$$

Exercise 5.24 *Let X be the process with cooperative branching defined above and set*

$$X_t^{(2)}(i) := 1_{\{X_t(i)=1=X_t(i+1)\}} \qquad (i \in \mathbb{Z},\ t \geq 0).$$

Show that X can be coupled to a contact process with double deaths Y (with the same parameter λ) in such a way that

$$Y_0 \leq X_0^{(2)} \quad \textit{implies} \quad Y_t \leq X_t^{(2)} \quad (t \geq 0).$$

Exercise 5.25 *Show that a system $(X_t)_{t\geq 0}$ of annihilating random walks can be coupled to a system $(Y_t)_{t\geq 0}$ of coalescing random walks such that*

$$X_0 \leq Y_0 \quad \textit{implies} \quad X_t \leq Y_t \quad (t \geq 0).$$

Note that the annihilating random walks are not a monotone particle system.

Exercise 5.26 *Let X be a system of branching and coalescing random walks with generator*

$$\begin{aligned} G_X f(x) = \tfrac{1}{2} b \sum_{i\in\mathbb{Z}} \sum_{\sigma\in\{-1,+1\}} \big\{f\big(\mathtt{bra}_{i,i+\sigma} x\big) - f\big(x\big)\big\} \\ + \tfrac{1}{2} \sum_{i\in\mathbb{Z}} \sum_{\sigma\in\{-1,+1\}} \big\{f\big(\mathtt{rw}_{i,i+\sigma} x\big) - f\big(x\big)\big\}, \end{aligned}$$

and let Y be a system of coalescing random walks with positive drift, with generator

$$\begin{aligned} G_Y f(y) = \tfrac{1}{2}(1 + b) \sum_{i\in\mathbb{Z}} \big\{f\big(\mathtt{rw}_{i,i+1} y\big) - f\big(y\big)\big\} \\ + \tfrac{1}{2} \sum_{i\in\mathbb{Z}} \big\{f\big(\mathtt{rw}_{i,i-1} y\big) - f\big(y\big)\big\}. \end{aligned}$$

Show that X and Y can be coupled such that

$$Y_0 \leq X_0 \quad \textit{implies} \quad Y_t \leq X_t \quad (t \geq 0).$$

Exercise 5.27 *Let $d < d'$ and identify $\mathbb{Z}^d$ with the subset of $\mathbb{Z}^{d'}$ consisting of all $(i_1, \ldots, i_{d'})$ with $(i_{d+1}, \ldots, i_{d'}) = (0, \ldots, 0)$. Let X and X′ denote the nearest-neighbor contact processes on $\mathbb{Z}^d$ and $\mathbb{Z}^{d'}$, respectively, with generator as in (1.8), with the same infection rate λ and death rate δ. Show that X and X′ can be coupled such that*

$$X_0(i) \leq X'_0(i) \quad (i \in \mathbb{Z}^d) \qquad \textit{implies} \qquad X_t(i) \leq X'_t(i) \quad (t \geq 0,\ i \in \mathbb{Z}^d).$$

Prove the same when X is the nearest-neighbor process, and X′ is the range R process (both on $\mathbb{Z}^d$). (Note that for these comparison arguments, the normalization in (1.8) is more convenient than the normalization in (5.7).)

6

Duality

6.1 Basic Definitions

Let S be a finite set, let Λ be countable, let $\mathcal{G}$ be a collection of local maps $m\colon S^\Lambda \to S^\Lambda$, and let $(r_m)_{m\in\mathcal{G}}$ be nonnegative rates satisfying (4.11). Then Theorem 4.14 tells us how the interacting particle system with generator

$$Gf(x) := \sum_{m\in\mathcal{G}} r_m\{f(m(x)) - f(x)\} \qquad (x \in S^\Lambda). \tag{6.1}$$

can be constructed from a graphical representation ω, which is a Poisson point set on $\mathcal{G}\times\mathbb{R}$ with intensity as in (4.8). More precisely, in (4.12), we have seen how in terms of ω it is possible to define a stationary stochastic flow $(\mathbf{X}_{s,t})_{s\leq t}$ with independent increments, so that if $s \in \mathbb{R}$ and X_0 is an S^Λ-valued random variable, independent of ω, then setting

$$X_t := \mathbf{X}_{s,s+t}(X_0) \qquad (t \geq 0) \tag{6.2}$$

defines a Feller process $(X_t)_{t\geq 0}$ whose generator is (the closure of) G defined in (4.1). We call this the interacting particle system with generator G.

Key to the proof of Theorem 4.14 was the backward in time process. For any finite set T, by (4.25), setting

$$\mathbf{F}_{t,s}(\phi) := \phi \circ \mathbf{X}_{s,t} \qquad \big(s \leq t,\ \phi \in C(S^\Lambda, T)\big), \tag{6.3}$$

defines a backward stochastic flow $(\mathbf{F}_{t,s})_{t\geq s}$ on the countable set $C(S^\Lambda, T)$. Alternatively[1] $(\mathbf{F}_{t,s})_{t\geq s}$ can be defined directly in terms of the graphical representation ω as in (4.20). If $u \in \mathbb{R}$ and Φ_0 is a random variable with values in $C(S^\Lambda, T)$, independent of ω, then setting

$$\Phi_t := \mathbf{F}_{u,u-t}(\Phi_0) \qquad (t \geq 0) \tag{6.4}$$

defines a nonexplosive continuous-time Markov chain $(\Phi_t)_{t\geq 0}$ with values in $C(S^\Lambda, T)$ and generator H as in (4.17). We called this the *backward in time process*. Due to the reversal of time, this Markov process has, somewhat unusually, left-continuous sample paths.

As we will see in the present chapter, the backward in time process is not just a useful tool in the construction of interacting particle systems but also in their further study. In fact, many of the most tractable and most studied interacting particle systems (such as the voter model and the contact process) are tractable precisely because their backward in time process is of a simple nature. The study of the backward in time process naturally leads to Markov process duality, which is the topic of the present chapter.

[1] In fact, in Chapter 4, we first defined $(\mathbf{F}_{t,s})_{t\geq s}$ by (4.20) and then used it to prove the existence of a forward stochastic flow $(\mathbf{X}_{s,t})_{s\leq t}$ as in Theorem 4.14 such that (6.3) holds.

Let S, R, and T be sets, and let $\psi\colon S\times R\to T$ be a function. Then we say that two maps $m\colon S\to S$ and $\hat m\colon R\to R$ are *dual* to each other with respect to the *duality function* ψ if

$$\psi\big(m(x),y\big)=\psi\big(x,\hat m(y)\big)\qquad (x\in S,\ y\in R).$$

If $(\mathbf{X}_{s,t})_{s\le t}$ is a stochastic flow (in the sense of (2.25)) on S and $(\mathbf{Y}_{t,s})_{t\ge s}$ is a backward stochastic flow (in the sense of (2.29)) on R, then we say that $(\mathbf{X}_{s,t})_{s\le t}$ and $(\mathbf{Y}_{t,s})_{t\ge s}$ are *dual* to each other with respect to the duality function ψ if

$$\psi\big(\mathbf{X}_{s,t}(x),y\big)=\psi\big(x,\mathbf{Y}_{t,s}(y)\big)\qquad (s\le t,\ x\in S,\ y\in R). \tag{6.5}$$

Fix $s<u$, let X_0 and Y_0 be independent of each other and of the stochastic flows $(\mathbf{X}_{s,t})_{s\le t}$ and $(\mathbf{Y}_{t,s})_{t\ge s}$, and let $(X_t)_{t\ge 0}$ and $(Y_t)_{t\ge 0}$ be the Markov processes defined by

$$X_t:=\mathbf{X}_{s,s+t}(X_0)\quad\text{and}\quad Y_t:=\mathbf{Y}_{u,u-t}(Y_0)\qquad (t\ge 0).$$

Then we claim that

$$\text{the function } [s,u]\ni t\mapsto \psi(X_{t-s},Y_{u-t}) \text{ is constant.} \tag{6.6}$$

Indeed, the duality of $(\mathbf{X}_{s,t})_{s\le t}$ and $(\mathbf{Y}_{t,s})_{t\ge s}$ implies

$$\begin{aligned}\psi(X_{t-s},Y_{u-t})&=\psi\big(\mathbf{X}_{s,t}(X_0),\mathbf{Y}_{u,t}(Y_0)\big)\\&=\psi\big(\mathbf{X}_{t,u}\circ\mathbf{X}_{s,t}(X_0),Y_0\big)=\psi\big(\mathbf{X}_{s,u}(X_0),Y_0\big),\end{aligned}$$

which clearly does not depend on t. A relation of the form (6.6) is called a *pathwise duality*.[2] In particular, setting $t=s,u$ in (6.6), we see that

$$\psi(X_u,Y_0)=\psi(X_0,Y_u).$$

In the special case that ψ takes values in a linear space such as $\mathbb{R}$ or $\mathbb{C}$, we can take expectations and conclude that

$$\mathbb{E}\big[\psi(X_u,Y_0)\big]=\mathbb{E}\big[\psi(X_0,Y_u)\big]\qquad (u\ge 0), \tag{6.7}$$

whenever X_u is independent of Y_0 and X_0 is independent of Y_u. A relation of the form (6.7) is called a *duality* between the Markov processes $(X_t)_{t\ge 0}$ and $(Y_t)_{t\ge 0}$.

We claim that any interacting particle system of the type described at the beginning of this section trivially has at least one pathwise dual, which is the backward in time process. To see this, fix a finite set T and let $\psi\colon S^\Lambda\times C(S^\Lambda,T)\to T$ be the duality function defined by

$$\psi(x,\phi):=\phi(x)\qquad \big(x\in S^\Lambda,\ \phi\in C(S^\Lambda,T)\big). \tag{6.8}$$

Then it is straightforward to check that the stochastic flow $(\mathbf{X}_{s,t})_{s\le t}$ and the backward stochastic flow $(\mathbf{F}_{t,s})_{t\ge s}$ are dual with respect to the duality function ψ, and hence the interacting particle system $(X_t)_{t\ge 0}$ and the backward in time process $(\Phi_t)_{t\ge 0}$ are pathwise dual.

Although formally, we have now found a pathwise dual for each interacting particle system, this dual is of little use in practice since the backward in time process is in general very complicated and the space $C(S^\Lambda,T)$ is very large. As we will see in the coming sections, however, it sometimes happens that $C(S^\Lambda,T)$ contains a subspace of "nice" functions that is

[2] This terminology was first introduced in [JK14b].

mapped into itself under the backward stochastic flow $(\mathbf{F}_{t,s})_{t\geq s}$, and this then leads to a more useful pathwise duality.

6.2 Additive Systems

There exists a useful duality theory for additive systems, which was first developed by Harris [Har78] and Griffeath [Gri79]. We first discuss additive systems and then in the next section discuss their duality. In line with notation introduced in Section 2.9, but specialized to $S = \{0, 1\}$, we set

$$\mathcal{S}(\Lambda) := \{0,1\}^\Lambda \quad \text{and} \quad \mathcal{S}_{\text{fin}}(\Lambda) := \big\{x \in \mathcal{S}(\Lambda) : |x| < \infty\big\},$$

where

$$|x| := \sum_{i\in\Lambda} x(i) \qquad \big(x \in \mathcal{S}(\Lambda)\big).$$

We equip $\mathcal{S}(\Lambda) := \{0,1\}^\Lambda$ with the product topology and $\mathcal{S}_{\text{fin}}(\Lambda)$, which is countable, with the discrete topology. As we have done before, we let $\underline{0}$ denote the configuration that is identically zero. We let $(x \vee y)(i) := x(i) \vee y(i)$ denote the pointwise maximum of two configurations $x, y \in \mathcal{S}(\Lambda)$. Let Λ, Δ be countable sets. By definition, a map $m\colon \mathcal{S}(\Lambda) \to \mathcal{S}(\Delta)$ is *additive* if and only if

$$\begin{aligned}
&\text{(i) } m(\underline{0}) = \underline{0},\\
&\text{(ii) } m(x \vee y) = m(x) \vee m(y) \qquad \big(x, y \in \mathcal{S}(\Lambda)\big).
\end{aligned}$$

The same definition applies to maps $m\colon \mathcal{S}_{\text{fin}}(\Lambda) \to \mathcal{S}(\Delta)$, where now (ii) needs to hold only for $x, y \in \mathcal{S}_{\text{fin}}(\Lambda)$. If m is moreover continuous, then

$$m\Big(\bigvee_{k=1}^{\infty} x_k\Big) = \bigvee_{k=1}^{\infty} m(x_k) \qquad \big(x_k \in \mathcal{S}(\Lambda)\big), \tag{6.9}$$

as follows by writing $m\Big(\lim_{n\to\infty} \bigvee_{k=1}^{n} x_k\Big) = \lim_{n\to\infty} m\Big(\bigvee_{k=1}^{n} x_k\Big) = \lim_{n\to\infty} \bigvee_{k=1}^{n} m(x_k)$. Each additive map is monotone, as follows by observing that $y \geq x$ implies $m(y) = m(x \vee y) = m(x) \vee m(y) \geq m(x)$. A lot of local maps we have already seen are additive. Examples are:

- The voter map $\texttt{vot}_{ij}$ defined in (1.4).
- The branching map $\texttt{bra}_{ij}$ defined in (1.6).
- The death map $\texttt{death}_i$ defined in (1.7).
- The exclusion map $\texttt{excl}_{ij}$ defined in (1.22).
- The coalescing random walk map $\texttt{rw}_{ij}$ defined in (1.24).

On the other hand, the following local maps are monotone but not additive:

- The asymmetric exclusion map $\texttt{asep}_{ij}$ defined in (1.20).
- The cooperative branching map $\texttt{coop}_{ijk}$ defined in (1.27).
- The maps $m_{i,L}$ defined in (4.47) to construct the Ising model with Glauber dynamics.

An interacting particle system is called *additive* if its generator can be represented in additive local maps. Examples of additive particle systems are:

- The voter model with generator as in (1.5).
- The contact process with generator as in (1.8).
- The biased voter model with generator as in (1.17).
- The symmetric exclusion process with generator as in (1.23).
- Systems of coalescing random walks with generator as in (1.25).

We define $e_i \in \mathcal{S}(\Lambda)$ by $e_i(j) := 1$ if $i = j$ and $:= 0$ otherwise. There is a useful graphical way to describe a continuous additive map $m\colon \mathcal{S}(\Lambda) \to \mathcal{S}(\Lambda)$, that works as follows:

- For each $i, j \in \Lambda$ with $i \neq j$ such that $m(e_i)(j) = 1$, we draw an arrow from i to j.
- For each $i \in \Lambda$ such that $m(e_i)(i) = 0$, we draw a blocking symbol ▬ at i.

The following lemma says that continuous additive maps are fully described by their arrows and blocking symbols.

Lemma 6.1 (Graphical description) *Let $m\colon \mathcal{S}(\Lambda) \to \mathcal{S}(\Lambda)$ be a continuous additive map, and let $x \in \{0,1\}^\Lambda$. Then $m(x)(j) = 1$ if and only if at least one of the following conditions is satisfied:*

(i) *for some $i \in \Lambda\backslash\{j\}$ with $x(i) = 1$, there is an arrow from i to j,*
(ii) *$x(j) = 1$, and there is no blocking symbol at j.*

Proof This follows from (6.9) by writing

$$m(x) = m\Big(\bigvee_{i:\,x(i)=1} e_i\Big) = \bigvee_{i:\,x(i)=1} m(e_i).$$

∎

In terms of arrows and blocking symbols, the branching map $\mathtt{bra}_{ij}$, the death map $\mathtt{death}_i$, the voter map $\mathtt{vot}_{ij}$, the coalescing random walk map $\mathtt{rw}_{ij}$, and the exclusion map $\mathtt{excl}_{ij}$ look as follows:

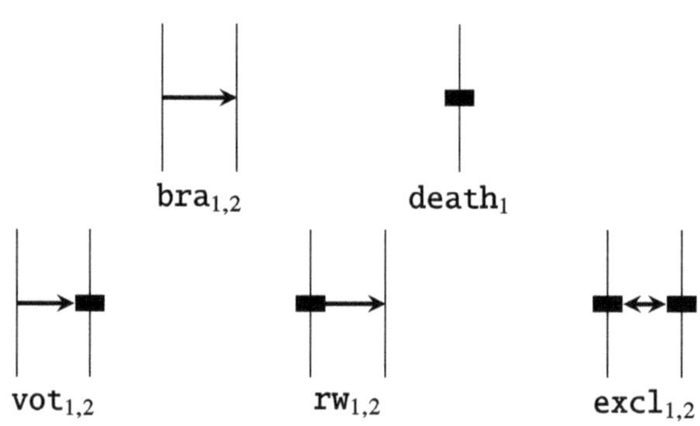

(6.10)

We use our conventions of representing additive maps in terms of arrows and blocking symbols to depict the graphical representation of an additive interacting particle system in a more suggestive way. In Figure 4.1, we drew the graphical representation of a contact process as in Figure 6.1.

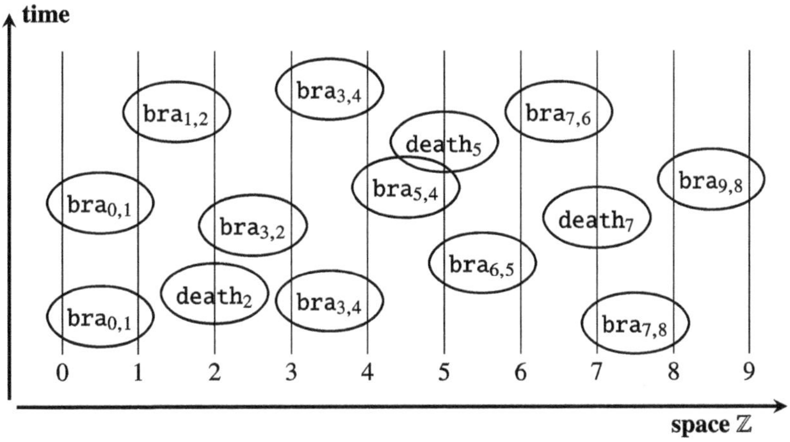

Figure 6.1 Graphical representation of a contact process with maps.

With our new conventions, the same graphical representation looks as in Figure 6.2.

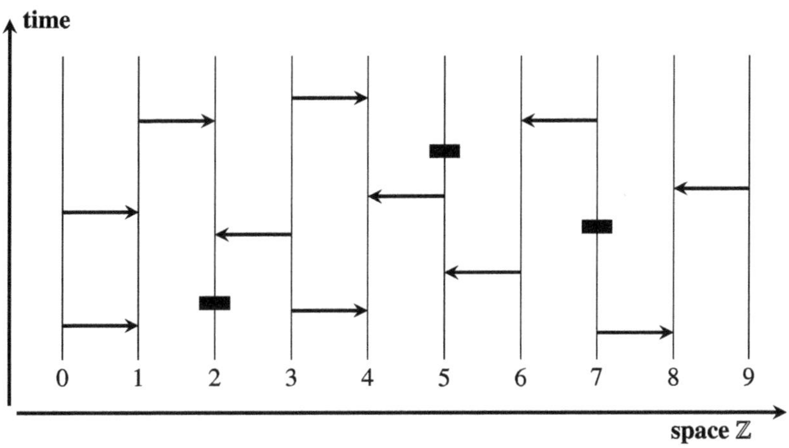

Figure 6.2 Graphical representation of a contact process with arrows and blocking symbols.

It is easy to see that the concatenation of two additive maps is again additive. As a result, using Proposition 4.23, we see that if $(\mathbf{X}_{s,t})_{s\leq t}$ is the stochastic flow associated with the graphical representation of an additive particle system, then the functions $\mathbf{X}_{s,t} : \{0,1\}^\Lambda \to \{0,1\}^\Lambda$ are additive maps. By Theorem 4.14, they are also continuous. We claim that $\mathbf{X}_{s,t}(x)(j) = 1$ if and only if there is an $i \in \Lambda$ such that $x(i) = 1$ and it is possible to walk through the graphical representation from the space-time point (i, s) to the space-time point (j, t) along an upward path that may use arrows but must avoid the blocking symbols. We now make this claim more precise.

For any $i, j \in \Lambda$ and $s < u$, by definition, an *open path* from (i, s) to (j, u) is a cadlag function $\gamma\colon [s,u] \to \Lambda$ such that $\gamma_s = i$, $\gamma_u = j$, and

$$\begin{array}{ll}\text{(i)} & \text{if } \gamma_{t-} \neq \gamma_t \text{ for some } t \in (s,u], \text{ then there is an} \\ & \text{arrow from } (\gamma_{t-},t) \text{ to } (\gamma_t,t), \\ \text{(ii)} & \text{there exist no } t \in (s,u] \text{ such that } \gamma_{t-} = \gamma_t \\ & \text{while there is a blocking symbol at } (\gamma_t,t).\end{array} \tag{6.11}$$

We write $(i,s) \rightsquigarrow (j,u)$ if there exists an open path from (i,s) to (j,u). With these definitions, we can make our earlier claim precise. We claim that:

$$\mathbf{X}_{s,t}(x)(j) = 1 \quad \text{if and only if} \quad \exists i \in \Lambda \text{ s.t. } x(i) = 1 \text{ and } (i,s) \rightsquigarrow (j,t). \tag{6.12}$$

To prove (6.12), it suffices to observe that if we define $X_t(j) := 1$ if and only if the condition on the right-hand side of (6.12) is satisfied, then the function $(X_t)_{t\geq s}$ solves the evolution equation (4.10). For example, for the graphical representation of the contact process that we earlier used as an example, the time evolution of the process $X_t := \mathbf{X}_{0,t}(X_0)$ $(t \geq 0)$ might look as in Figure 6.3.

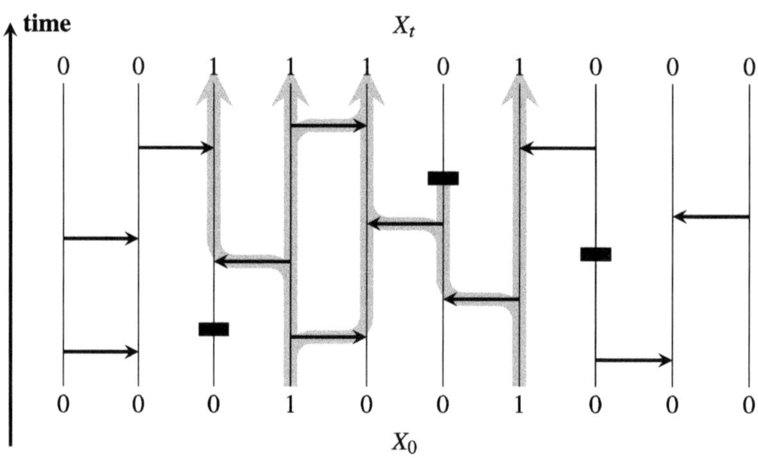

Figure 6.3 Contact process defined by open paths.

Recall that if E is any topological space, then a function $f\colon E \to \mathbb{R}$ is called *lower semicontinuous* if one, and hence both, of the following equivalent conditions are satisfied:

(i) $\liminf_{n\to\infty} f(x_n) \geq f(x)$ whenever $x_n \to x$,
(ii) the level set $\{x \in E : f(x) \leq a\}$ is closed for each $a \in \mathbb{R}$.

Exercise 6.2 *Show that (6.9) remains true if m is additive and lower semicontinuous.*

6.3 Additive Duality

Recall the definition of $\mathcal{S}_{\text{fin}}(\Lambda)$ from the beginning of the previous section. For countable sets Λ, Δ, we let

$$\mathcal{C}_{\text{add}}\big(\mathcal{S}(\Lambda), \mathcal{S}(\Delta)\big) \quad \text{and} \quad \mathcal{C}_{\text{add}}\big(\mathcal{S}_{\text{fin}}(\Lambda), \mathcal{S}_{\text{fin}}(\Delta)\big)$$

denote the space of continuous additive maps $m\colon \mathcal{S}(\Lambda) \to \mathcal{S}(\Delta)$ and the space of additive maps $m\colon \mathcal{S}_{\rm fin}(\Lambda) \to \mathcal{S}_{\rm fin}(\Delta)$, respectively. Since $\mathcal{S}_{\rm fin}(\Lambda)$ is equipped with the discrete topology, the latter are automatically continuous. We define a function $\psi_{\rm add}\colon \mathcal{S}(\Lambda) \times \mathcal{S}(\Lambda) \to \{0,1\}$ by

$$\psi_{\rm add}(x,y) := 1_{\{x \wedge y \neq \underline{0}\}} \qquad \big(x, y \in \mathcal{S}(\Lambda)\big), \tag{6.13}$$

where $x \wedge y$ denotes the pointwise minimum of x and y, and $\underline{0}$ denotes the all zero configuration. We adopt the following notation. For each $y \in \mathcal{S}(\Lambda)$, we define $\psi_{\rm add}(\,\cdot\,, y)\colon \mathcal{S}(\Lambda) \to \{0,1\}$ by

$$\psi_{\rm add}(\,\cdot\,, y)(x) := \psi_{\rm add}(x,y) \qquad \big(x \in \mathcal{S}(\Lambda)\big),$$

and we let $\psi_{\rm add}(\,*\,, y)$ denote the restriction of $\psi_{\rm add}(\,\cdot\,, y)$ to $\mathcal{S}_{\rm fin}(\Lambda)$. We define $\psi_{\rm add}(x, \,\cdot\,)$ and $\psi_{\rm add}(x, \,*\,)$ in the same way.

Lemma 6.3 (Additive duality function) *One has*

$$\begin{aligned} &\text{(i)} \quad C_{\rm add}\big(\mathcal{S}(\Lambda), \{0,1\}\big) = \big\{\psi_{\rm add}(\,\cdot\,, y) : y \in \mathcal{S}_{\rm fin}(\Lambda)\big\},\\ &\text{(ii)} \quad C_{\rm add}\big(\mathcal{S}_{\rm fin}(\Lambda), \{0,1\}\big) = \big\{\psi_{\rm add}(x, \,*\,) : x \in \mathcal{S}(\Lambda)\big\}. \end{aligned} \tag{6.14}$$

Moreover, $y \mapsto \psi_{\rm add}(\,\cdot\,, y)$ is a bijection from $\mathcal{S}_{\rm fin}(\Lambda)$ to $C_{\rm add}\big(\mathcal{S}(\Lambda), \{0,1\}\big)$ and $x \mapsto \psi_{\rm add}(x, \,*\,)$ is a bijection from $\mathcal{S}(\Lambda)$ to $C_{\rm add}\big(\mathcal{S}_{\rm fin}(\Lambda), \{0,1\}\big)$.

Proof It is straightforward to check that $\mathcal{S}(\Lambda) \ni x \mapsto \psi_{\rm add}(x,y) \in \{0,1\}$ is additive for each $y \in \mathcal{S}(\Lambda)$ and by symmetry an analogue statement holds for $y \mapsto \psi_{\rm add}(x,y)$. It follows from Lemma 2.30 that $\psi_{\rm add}(\,\cdot\,, y)$ is continuous if $y \in \mathcal{S}_{\rm fin}(\Lambda)$. This proves the inclusions $\supset$ in (6.14) (i) and (ii).

To prove the converse inclusion in (6.14) (i), assume that $\phi\colon \mathcal{S}(\Lambda) \to \{0,1\}$ is continuous and additive. Define $y \in \mathcal{S}(\Lambda)$ by $y(i) := 1$ if $\phi(e_i) = 1$ and $:= 0$ otherwise. Since $\phi(\underline{0}) = 0$, we have $\phi(e_i) = 0$ for all $i \in \Lambda\backslash\mathcal{R}(\phi)$, and hence $y \in \mathcal{S}_{\rm fin}(\Lambda)$ by the continuity of ϕ. Now (6.9) gives

$$\phi(x) = \phi\Big(\bigvee_{i:\, x(i)=1} e_i\Big) = \bigvee_{i:\, x(i)=1} \phi(e_i) = \psi_{\rm add}(x,y).$$

The proof of the inclusion $\subset$ in (6.14) (ii) is similar. In this case, we can't use (6.9), but we don't need to since $y \in \mathcal{S}_{\rm fin}(\Lambda)$. To see that $y \mapsto \psi_{\rm add}(\,\cdot\,, y)$ and $x \mapsto \psi_{\rm add}(x, \,*\,)$ are bijections, it suffices to note that if $x(i) \neq x'(i)$, then $\psi_{\rm add}(x, e_i) \neq \psi_{\rm add}(x', e_i)$. ∎

We now consider an interacting particle system whose generator G has a random mapping representation of the form (6.1). We assume that $S = \{0,1\}$, and that all maps $m \in \mathcal{G}$ are additive. We also assume that the rates satisfy (4.11) so that the stochastic flow $(\mathbf{X}_{s,t})_{s\leq t}$ and the backward stochastic flow $(\mathbf{F}_{t,s})_{t\geq s}$ are well-defined. Since the concatenation of two additive functions is again additive, we have that

$$\phi \in C_{\rm add}(S^\Lambda, \{0,1\}) \quad \text{implies} \quad \mathbf{F}_{t,s}(\phi) \in C_{\rm add}(S^\Lambda, \{0,1\}) \qquad (t \geq s).$$

By Lemma 6.3, there is a one-to-one correspondence between functions $\phi \in C_{\rm add}(S^\Lambda, \{0,1\})$ and configurations $y \in \mathcal{S}_{\rm fin}(\Lambda)$. It follows that we can define a backward stochastic flow $(\mathbf{Y}_{t,s})_{t\geq s}$ on $\mathcal{S}_{\rm fin}(\Lambda)$ by

$$\mathbf{F}_{t,s}(\psi_{\rm add}(\,\cdot\,, y)) =: \psi_{\rm add}\big(\,\cdot\,, \mathbf{Y}_{t,s}(y)\big) \qquad \big(t \geq s,\ y \in \mathcal{S}_{\rm fin}(\Lambda)\big). \tag{6.15}$$

We will show that $(\mathbf{Y}_{t,s})_{t\geq s}$ can be used to define a Markov process $(Y_t)_{t\geq 0}$ that is itself an additive particle system and that is pathwise dual to the system $(X_t)_{t\geq 0}$ with generator G. We first state the main facts and then prove them.

Lemma 6.4 (Dual maps) *For each local additive map $m\colon \mathcal{S}(\Lambda) \to \mathcal{S}(\Lambda)$, there exists a unique map $\hat{m}\colon \mathcal{S}(\Lambda) \to \mathcal{S}(\Lambda)$ that is dual to m with respect to the duality function ψ_{add}, in the sense that*

$$\psi_{\mathrm{add}}\big(m(x), y\big) = \psi_{\mathrm{add}}\big(x, \hat{m}(y)\big) \qquad \big(x, y \in \mathcal{S}(\Lambda)\big). \tag{6.16}$$

This dual map is also local and additive and uniquely characterized by

$$m(e_i)(j) = 1 \quad \Leftrightarrow \quad \hat{m}(e_j)(i) = 1 \qquad (i, j \in \Lambda). \tag{6.17}$$

We observe that in terms of our graphical way of depicting additive maps, formula (6.17) has the following interpretation:

$\hat{m}$ is obtained from m by keeping the blocking symbols and reversing the direction of all arrows. (6.18)

This means that the duals of the maps depicted below Lemma 6.1 are given by:

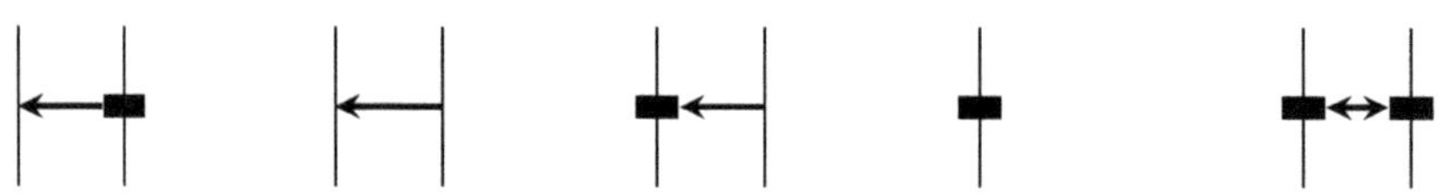

$$\widehat{\mathtt{vot}}_{1,2} \qquad \widehat{\mathtt{bra}}_{1,2} \qquad \widehat{\mathtt{rw}}_{1,2} \qquad \widehat{\mathtt{death}}_{1} \qquad \widehat{\mathtt{excl}}_{1,2} \tag{6.19}$$

We see from this that:

$$\begin{aligned} \widehat{\mathtt{bra}}_{ij} &= \mathtt{bra}_{ji}, \quad \widehat{\mathtt{death}}_i = \mathtt{death}_i, \\ \widehat{\mathtt{vot}}_{ij} = \mathtt{rw}_{ji}, \quad \widehat{\mathtt{rw}}_{ij} &= \mathtt{vot}_{ji}, \quad \widehat{\mathtt{excl}}_{ij} = \mathtt{excl}_{ij}. \end{aligned} \tag{6.20}$$

Theorem 6.5 (Additive duality) *Let G be the generator of an interacting particle system $(X_t)_{t\geq 0}$ with state space $\mathcal{S}(\Lambda)$. Assume that G has a random mapping representation of the form (6.1) such that all local maps $m \in \mathcal{G}$ are additive and the rates $(r_m)_{m\in\mathcal{G}}$ satisfy (4.11). Then*

$$\hat{G} := \sum_{m\in\mathcal{G}} r_m\big\{f\big(\hat{m}(y)\big) - f(y)\big\} \qquad \big(y \in \mathcal{S}_{\mathrm{fin}}(\Lambda)\big) \tag{6.21}$$

is the generator of a nonexplosive continuous-time Markov chain $(Y_t)_{t\geq 0}$ with state space $\mathcal{S}_{\mathrm{fin}}(\Lambda)$. Let ω be a graphical representation associated with the random mapping representation (6.1) of G. Define a graphical representation $\hat{\omega}$ associated with the random mapping representation (6.21) of $\hat{G}$ by

$$\hat{\omega} := \big\{(\hat{m}, t) : (m, t) \in \omega\big\}.$$

Let $(\mathbf{X}_{s,t})_{s\leq t}$ be the stochastic flow on $\mathcal{S}(\Lambda)$ defined in terms of ω as in Theorem 4.14, and let $(\mathbf{Y}_{t,s})_{t\geq s}$ be the backward stochastic flow on $\mathcal{S}_{\mathrm{fin}}(\Lambda)$ defined in terms of $\hat{\omega}$ as in Theorem 2.24. Then almost surely

$$\psi_{\mathrm{add}}\big(\mathbf{X}_{s,t}(x), y\big) = \psi_{\mathrm{add}}\big(x, \mathbf{Y}_{t,s}(y)\big) \qquad \big(s \leq t,\ x \in \mathcal{S}(\Lambda),\ y \in \mathcal{S}_{\mathrm{fin}}(\Lambda)\big). \tag{6.22}$$

If the random mapping representation (6.21) also satisfies (4.11) so that $(\mathbf{Y}_{t,s})_{t\geq s}$ can be extended to $\mathcal{S}(\Lambda)$, then (6.22) holds for all $x, y \in \mathcal{S}(\Lambda)$.

Formula (6.22) says that the interacting particle systems with generators G and $\hat{G}$ are pathwise dual with duality function ψ_{add}; see the discussion in Section 6.1. This duality has a simple graphical interpretation. We recall from (6.18) that we can obtain the dual of an additive local map by reversing the direction of all arrows and keeping all blocking symbols. In Section 6.2, we constructed a contact process $(X_t)_{t\geq 0}$ from its graphical representation in terms of open paths, as in Figure 6.4.

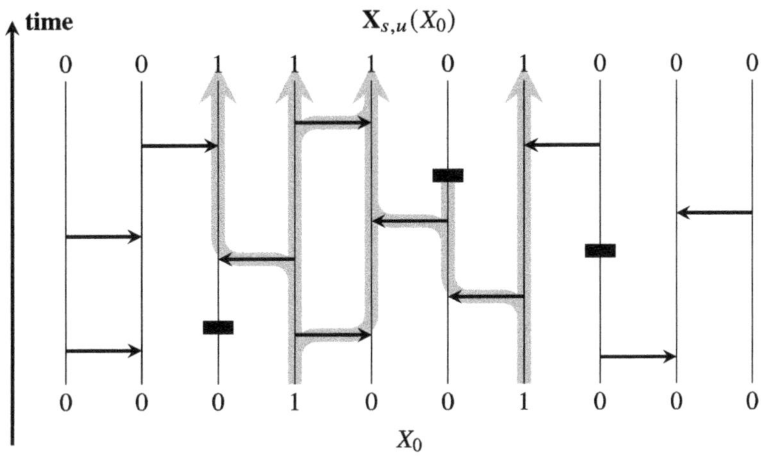

Figure 6.4 Contact process defined by open paths.

Using the recipe "reverse the arrows, keep the blocking symbols", we can construct the dual process $(Y_t)_{t\geq 0}$ as in Figure 6.5.

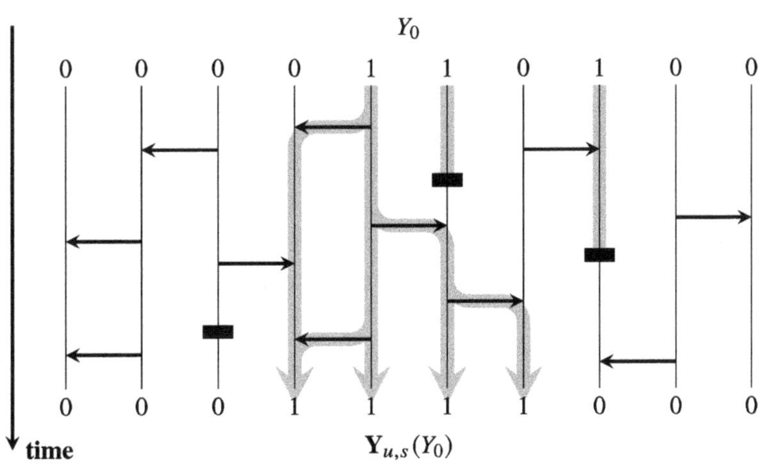

Figure 6.5 Dual contact process defined by open paths.

The duality relation between the forward and backward stochastic flows then follows from the observation that

$$
\begin{aligned}
&\psi_{\text{add}}\big(\mathbf{X}_{s,u}(x), y\big) = 1 \\
&\quad \Leftrightarrow \exists i, j \in \Lambda \text{ s.t. } x(i) = 1,\ y(j) = 1,\ (i,s) \rightsquigarrow (j,u) \\
&\quad \Leftrightarrow \psi_{\text{add}}\big(x, \mathbf{Y}_{u,s}(y)\big) = 1.
\end{aligned}
$$

In our previous example of the contact process, the dual process is also a contact process, but in general, the dual process can have a different dynamics from the forward in time process. For example, we see from (6.20) that the additive dual of the voter model is a system of coalescing random walks.

We still need to prove Lemma 6.4 and Theorem 6.5.

Proof of Lemma 6.4 If $m(e_i)(j) = 1$ for some $i \neq j$, then by the fact that $m(\underline{0}) = 0$ we see that $j \in \mathcal{D}(m)$ and $i \in \mathcal{R}(m[j])$. Also, if $m(e_i)(i) = 0$, then $i \in \mathcal{D}(m)$. It follows that the graphical representation of an additive local map contains only finitely many arrows and blocking symbols. Setting $x = e_i$ and $y = e_j$ in (6.16) gives (6.17), so the latter is clearly necessary. Using the recipe "reverse the arrows, keep the blocking symbols", we can find a local map $\hat{m}$ such that (6.17) holds. Using (6.9), which is applicable since both m and $\hat{m}$ are continuous, we observe that

$$
\psi_{\text{add}}\big(m(x), y\big) = \psi_{\text{add}}\Big(\bigvee_{i:\, x(i)=1} m(e_i), y\Big) = \bigvee_{i:\, x(i)=1}\ \bigvee_{j:\, y(j)=1} 1_{\{m(e_i)(j) = 1\}},
$$

which by (6.17) and the same argument backwards is equal to $\psi_{\text{add}}\big(x, \hat{m}(y)\big)$. Since $\hat{m}(y)(i) = \psi_{\text{add}}(e_i, \hat{m}(y)) = \psi_{\text{add}}(m(e_i), y)$, we see that $\hat{m}$ is the unique map from $\mathcal{S}(\Lambda)$ into itself that is dual to m with respect to ψ_{add}. ∎

Proof of Theorem 6.5 Under the condition (4.11), it has been shown in Proposition 4.16 that

$$
Hf(\phi) := \sum_{m \in \mathcal{G}} r_m \big\{ f(\phi \circ m) - f(\phi) \big\}
$$

is the generator of a nonexplosive continuous-time Markov chain, called the backward in time process, with state space $C(\mathcal{S}(\Lambda), \{0,1\})$. In (4.20), we used the graphical representation ω to define a backward stochastic flow $(\mathbf{F}_{t,s})_{t \ge s}$ associated with this backward in time process. Since the composition of two additive maps is additive, this backward stochastic flow maps the space $C_{\text{add}}(\mathcal{S}(\Lambda), \{0,1\})$ into itself. By Lemma 6.3, $y \mapsto C_{\text{add}}(\mathcal{S}(\Lambda), \{0,1\})$ is a bijection from $\mathcal{S}_{\text{fin}}(\Lambda)$ to $C_{\text{add}}(\mathcal{S}(\Lambda), \{0,1\})$, which allows us to define $(\mathbf{Y}_{t,s})_{t \ge s}$ in terms of $(\mathbf{F}_{t,s})_{t \ge s}$ as in (6.15). In view of (4.19), this means that for each $u \in \mathbb{R}$ and $y \in \mathcal{S}_{\text{fin}}(\Lambda)$, the function $(Y_t)_{t \le u}$ defined as $Y_t := \mathbf{Y}_{u,t}(y)$ $(t \le u)$ is the unique piecewise constant, right-continuous solution of the evolution equation

$$
Y_{t-} = \begin{cases} Y' \text{ where } \psi_{\text{add}}(\,\cdot\,, Y') := \psi_{\text{add}}(\,\cdot\,, Y_t) \circ m & \text{if } (m,t) \in \omega, \\ Y_t & \text{otherwise.} \end{cases}
$$

We observe that for any $x \in \mathcal{S}(\Lambda)$ and $y \in \mathcal{S}_{\text{fin}}(\Lambda)$,

$$
\psi_{\text{add}}(\,\cdot\,, y) \circ m(x) = \psi_{\text{add}}\big(m(x), y\big) = \psi_{\text{add}}\big(x, \hat{m}(y)\big),
$$

so $(\mathbf{Y}_{t,s})_{t\geq s}$ is the backward stochastic flow associated with the continuous-time Markov chain with generator $\hat{G}$ as in (6.21). In particular, the latter is nonexplosive since the backward in time process is. Now (6.3) and (6.15) imply that

$$\psi_{\text{add}}\big(\mathbf{X}_{s,t}(x), y\big) = \psi_{\text{add}}(\,\cdot\,, y) \circ \mathbf{X}_{s,t}(x) = \mathbf{F}_{t,s}\big(\psi_{\text{add}}(\,\cdot\,, y)\big)(x) = \psi_{\text{add}}\big(x, \mathbf{Y}_{t,s}(y)\big)$$

for all $t \geq s$, $x \in \mathcal{S}(\Lambda)$, and $y \in \mathcal{S}_{\text{fin}}(\Lambda)$, proving (6.22).

Assume that the random mapping representation (6.21) also satisfies (4.11) so that $(\mathbf{Y}_{t,s})_{t\geq s}$ can be extended to $\mathcal{S}(\Lambda)$. For each $y \in \mathcal{S}(\Lambda)$ we can find $y_n \in \mathcal{S}_{\text{fin}}(\Lambda)$ such that $y_n \uparrow y$. Then $\mathbf{Y}_{t,s}(y_n) \uparrow \mathbf{Y}_{t,s}(y)$ by the continuity of $\mathbf{Y}_{t,s}$ (proved in Theorem 4.14) and the monotonicity of $\mathbf{Y}_{t,s}$ (which follows from its additivity), so taking the limit in (6.22) we see that the latter holds for all $x, y \in \mathcal{S}(\Lambda)$. ∎

We conclude this section with a simple lemma that is important in applications of Theorem 6.5 and in particular of the duality relation (6.22). It says that the values of $\mathbb{E}[\psi_{\text{add}}(X_t, y)]$ for all $y \in \mathcal{S}_{\text{fin}}(\Lambda)$ determine the law of X_t uniquely.

Lemma 6.6 (Distribution determining functions) *The class of functions $\{\psi_{\text{add}}(\,\cdot\,, y) : y \in \mathcal{S}_{\text{fin}}(\Lambda)\}$ is distribution determining on $\mathcal{S}(\Lambda)$.*

Proof We may equivalently prove that the functions $g_y(x) := 1 - \psi_{\text{add}}(\,\cdot\,, y) = 1_{\{x\wedge y=\underline{0}\}}$ are distribution determining. Since $1_{\{x\wedge e_i=\underline{0}\}} = 1 - x(i)$, the class $\{g_y : y \in \mathcal{S}_{\text{fin}}(\Lambda)\}$ separates points, and since $g_y g_{y'} = g_{y\vee y'}$, this class is closed under products. The claim now follows from Lemma 4.39. ∎

Exercise 6.7 *Let $\mathcal{L}_{\text{add}}(\mathcal{S}(\Lambda), \{0, 1\})$ denote the space of lower semi-continuous additive maps $m \colon \mathcal{S}(\Lambda) \to \{0, 1\}$. In analogy with Lemma 6.3, show that*

$$\mathcal{L}_{\text{add}}\big(\mathcal{S}(\Lambda), \{0, 1\}\big) = \big\{\psi_{\text{add}}(\,\cdot\,, y) : y \in \mathcal{S}(\Lambda)\big\}.$$

Show that under the assumptions of Theorem 6.5, the backward stochastic flow $(\mathbf{F}_{t,s})_{t\geq s}$ maps the space $\mathcal{L}_{\text{add}}(\mathcal{S}(\Lambda), \{0, 1\})$ into itself. Use this to show that the definition of the backward stochastic flow $\mathbf{Y}_{t,s}(y)$ in (6.15) can be extended to $y \in \mathcal{S}(\Lambda)$, even if the random mapping representation (6.21) does not satisfy the summability condition (4.11).

Exercise 6.8 *Generalize Lemma 6.4 to lower semicontinuous additive maps by showing that each lower semicontinuous additive map $m \colon \mathcal{S}(\Lambda) \to \mathcal{S}(\Lambda)$ has a unique dual map $\hat{m}$ with respect to the duality function ψ_{add}, and that this dual map $\hat{m}$ is also lower semicontinuous and additive.*

Exercise 6.9 *Give an example of an additive particle system for which the forward generator G satisfies the summability condition (4.11), but the dual generator $\hat{G}$ does not.* Hint: *Consider a contact process and its dual contact process on a binary tree where for the forward process all infections point away from the root and for the dual process all infections point in the direction of the root.*

Exercise 6.10 *Let $(\beta_i)_{i\in\mathbb{N}}$ be positive constants, and consider the interacting particle system with lattice $\Lambda := \mathbb{N}$, local state space $S := \{0, 1\}$, and generator*

$$Gf(x) := \sum_{i=0}^{\infty} \beta_i \big\{f\big(\mathtt{bra}_{i+1,i}(x)\big) - f(x)\big\},$$

where $\texttt{bra}_{j,i}$ *is the branching map defined in (1.6). This process is additive, and the generator of the dual process is*

$$\hat{G}f(y) = \sum_{i=0}^{\infty} \beta_i \big\{ f\big(\texttt{bra}_{i,i+1}(y)\big) - f(y) \big\}.$$

If the rates β_i *are not bounded as a function of i, then (4.11) fails, so Theorem 4.14 is not applicable. Nevertheless, condition (4.21) is still satisfied, so by Lemma 4.17, the backward in time process is still a well-defined continuous-time Markov chain even though it may be explosive if* β *grows too fast as a function of i. (Compare Exercise 2.27.) In this case, by Theorem 2.22, we can still construct the backward stochastic flow* $(\mathbf{Y}_{t,s})_{t\geq s}$ *associated with the continuous-time Markov chain with generator* $\hat{G}$. *Here* $\mathbf{Y}_{t,s}$ *is a map from the space* $\mathcal{S}_{\rm fin}(\mathbb{N}) \cup \{\infty\}$ *into itself, where* $\mathbf{Y}_{t,s}(y) = \infty$ *means that the dual process started at time t and run backwards to time* $s < t$ *has exploded. We can use this to define a function* $X\colon [0,\infty) \to \{0,1\}^{\mathbb{N}}$ *by*

$$X_t(i) := \begin{cases} 0 & \text{if } \mathbf{Y}_{t,0}(e_i) \in \mathcal{S}_{\rm fin}(\mathbb{N}), \\ 1 & \text{if } \mathbf{Y}_{t,0}(e_i) = \infty. \end{cases} \qquad (i \in \mathbb{N},\ t \geq 0).$$

Show that $(X_t)_{t\geq 0}$ *solves the evolution equation (4.10) with initial state* $X_0 = \underline{0}$. *Since the all zero function is another solution, this shows that for this interacting particle system, solutions to the evolution equation (4.10) are not unique if the backward in time process is explosive.*

6.4 Cancellative Systems and Their Duals

The duality theory for cancellative systems was initiated by Griffeath [Gri79]. Let $\oplus$ denote addition modulo two, that is,

$$0 \oplus 0 := 0, \quad 0 \oplus 1 := 1, \quad 1 \oplus 0 := 1, \quad \text{and} \quad 1 \oplus 1 := 0.$$

Let Λ and Δ be countable sets. For $x, y \in \mathcal{S}(\Lambda)$, we define $(x \oplus y)(i) := x(i) \oplus y(i)$ $(i \in \Lambda)$ in a pointwise way. By definition, a map $m\colon \mathcal{S}(\Lambda) \to \mathcal{S}(\Delta)$ is *cancellative* if and only if

(i) $m(\underline{0}) = \underline{0}$,

(ii) $m(x \oplus y) = m(x) \oplus m(y) \qquad \big(x, y \in \mathcal{S}(\Lambda)\big)$.

The same definition applies to maps $m\colon \mathcal{S}_{\rm fin}(\Lambda) \to \mathcal{S}_{\rm fin}(\Delta)$, where in this case (ii) only needs to hold for $x, y \in \mathcal{S}_{\rm fin}(\Lambda)$. An interacting particle system is called *cancellative* if its generator can be represented in cancellative local maps. Examples of cancellative maps are:

- The voter map $\texttt{vot}_{ij}$ defined in (1.4).
- The death map $\texttt{death}_i$ defined in (1.7).
- The exclusion map $\texttt{excl}_{ij}$ defined in (1.22).
- The annihilating random walk map $\texttt{arw}_{ij}$ defined in (1.26).
- The annihilating branching map $\texttt{abra}_{ij}$ defined in (6.23).

Here, we define an *annihilating branching map* $\mathtt{abra}_{ij}\colon \{0,1\}^\Lambda \to \{0,1\}^\Lambda$ by

$$\mathtt{abra}_{ij}(x)(k) := \begin{cases} x(i) \oplus x(j) & \text{if } k = j, \\ x(k) & \text{otherwise.} \end{cases} \tag{6.23}$$

For countable sets Λ, Δ, we let

$$\mathcal{C}_{\rm canc}\big(\mathcal{S}(\Lambda), \mathcal{S}(\Delta)\big) \quad \text{and} \quad \mathcal{C}_{\rm canc}\big(\mathcal{S}_{\rm fin}(\Lambda), \mathcal{S}_{\rm fin}(\Delta)\big)$$

denote the space of continuous cancellative maps $m\colon \mathcal{S}(\Lambda) \to \mathcal{S}(\Delta)$ and the space of cancellative maps $m\colon \mathcal{S}_{\rm fin}(\Lambda) \to \mathcal{S}_{\rm fin}(\Delta)$, respectively. Similar to what we did in the additive case, we can describe a continuous cancellative map $m\colon \mathcal{S}(\Lambda) \to \mathcal{S}(\Lambda)$ in terms of arrows and blocking symbols:

- For each $i, j \in \Lambda$ with $i \neq j$ such that $m(e_i)(j) = 1$, we draw an arrow from i to j.
- For each $i \in \Lambda$ such that $m(e_i)(i) = 0$, we draw a blocking symbol ▬ at i.

The following lemma says that continuous cancellative maps are fully described by their arrows and blocking symbols.

Lemma 6.11 (Graphical description) *Let $m\colon \mathcal{S}(\Lambda) \to \mathcal{S}(\Lambda)$ be a continuous cancellative map, and let $x \in \{0,1\}^\Lambda$. For each $j \in \Lambda$, let R_j denote the set of $i \in \Lambda$ such that either $i = j$ and there is no blocking symbol at i, or $i \neq j$ and there is an arrow from i to j. Then R_j is finite and*

$$m(x)(j) = 1 \quad \Leftrightarrow \quad \big|\{i \in R_j : x(i) = 1\}\big| \text{ is odd.}$$

Proof Since $m(\underline{0}) = 0$, we see that $R_j \subset \mathcal{R}(m[j])$, where by the continuity of m, the latter is finite for each $j \in \Lambda$. Using the fact that we can change x outside $\mathcal{R}(m[j])$ without changing $m(x)(j)$, we see that

$$m(x)(j) = m[j]\Big(\bigoplus_{i \in \mathcal{R}(m[j]) : x(i)=1} e_i\Big) = \bigoplus_{i \in \mathcal{R}(m[j]) : x(i)=1} m[j](e_i),$$

which is one if and only if $\{i \in R_j : x(i) = 1\}$ has an odd number of elements. ∎

Every graphical representation involving arrows and blocking symbols that can be used to define an additive particle system can also be used to define a cancellative particle system. The cancellative maps mentioned at the beginning of this section have the following representations in terms of arrows and blocking symbols:

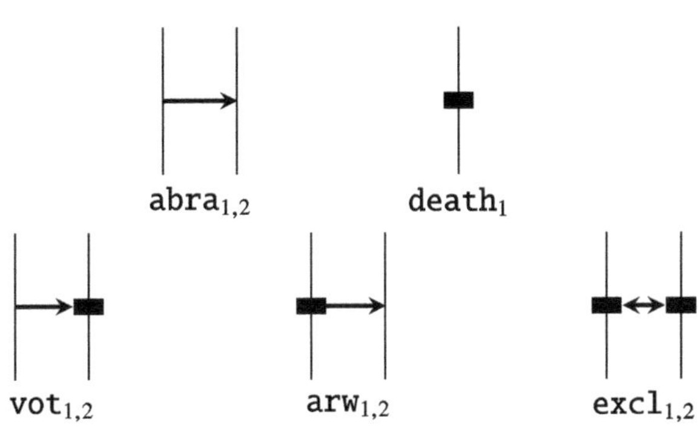

(6.24)

If we interpret the graphical representation of a contact process in a cancellative way, then it becomes a graphical representation for an interacting particle system involving the annihilating branching map $\mathtt{abra}_{ij}$ and the death map $\mathtt{death}_i$, see Figure 6.6. This system has been studied in [BDD91].

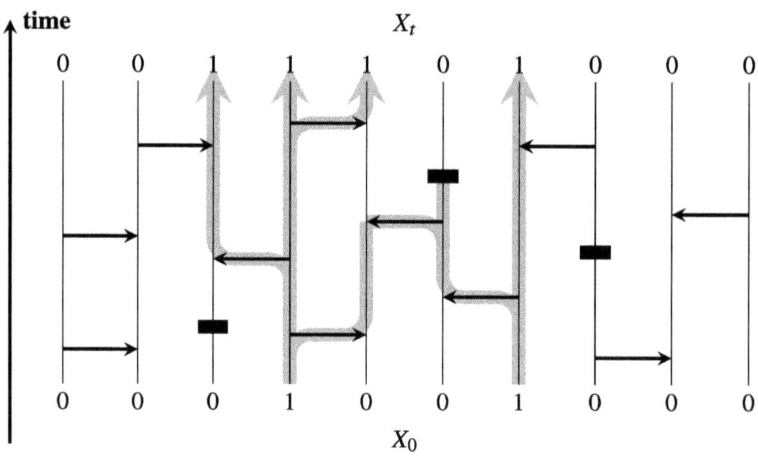

Figure 6.6 Graphical representation of a cancellative version of the contact process.

We define a *cancellative duality function* $\psi_{\rm canc}\colon \mathcal{S}(\Lambda)\times\mathcal{S}_{\rm fin}(\Lambda)\to\{0,1\}$ by

$$\psi_{\rm canc}(x,y) := \bigoplus_{i\in\Lambda} x(i)y(i) \qquad \big(x\in\mathcal{S}(\Lambda),\ y\in\mathcal{S}_{\rm fin}(\Lambda)\big). \tag{6.25}$$

Note that since $y\in\mathcal{S}_{\rm fin}(\Lambda)$, all but finitely many of the summands are zero, so the infinite sum modulo two is well-defined. Unlike in the additive case, there is no way to make sense of $\psi_{\rm canc}(x,y)$ for general $x,y\in\mathcal{S}(\Lambda)$.[3] The following lemma is similar to Lemma 6.3. Below, $\psi_{\rm add}(\,\cdot\,,y)$ denotes the map $\mathcal{S}(\Lambda)\ni x\mapsto\psi_{\rm add}(x,y)$, and $\psi_{\rm add}(\,*\,,y)$ denotes its restriction to $\mathcal{S}_{\rm fin}(\Lambda)$.

Lemma 6.12 (Cancellative duality function) *One has*

$$\begin{aligned} &\text{(i)} \quad C_{\rm canc}\big(\mathcal{S}(\Lambda),\{0,1\}\big) = \big\{\psi_{\rm canc}(\,\cdot\,,y) : y\in\mathcal{S}_{\rm fin}(\Lambda)\big\},\\ &\text{(ii)} \quad C_{\rm canc}\big(\mathcal{S}_{\rm fin}(\Lambda),\{0,1\}\big) = \big\{\psi_{\rm canc}(x,\,*\,) : x\in\mathcal{S}(\Lambda)\big\}. \end{aligned} \tag{6.26}$$

Moreover, $y\mapsto\psi_{\rm canc}(\,\cdot\,,y)$ *is a bijection from* $\mathcal{S}_{\rm fin}(\Lambda)$ *to* $C_{\rm canc}\big(\mathcal{S}(\Lambda),\{0,1\}\big)$ *and* $x\mapsto\psi_{\rm canc}(x,\,*\,)$ *is a bijection from* $\mathcal{S}(\Lambda)$ *to* $C_{\rm canc}\big(\mathcal{S}_{\rm fin}(\Lambda),\{0,1\}\big)$.

Proof The proof is almost the same as the proof of Lemma 6.3; only the proof of the inclusion $\subset$ in (6.26) (i) is a bit more complicated than in the additive case. To prove this inclusion, assume that $\phi\colon\mathcal{S}(\Lambda)\to\{0,1\}$ is continuous and cancellative. Define $y\in\mathcal{S}(\Lambda)$ by $y(i):=1$ if $\phi(e_i)=1$ and $:=0$ otherwise. Since $\phi(\underline{0})=0$, we have $\phi(e_i)=0$ for all $i\in\Lambda\backslash\mathcal{R}(\phi)$

[3] For interacting particle systems on $\Lambda=\mathbb{Z}$, it is sometimes useful to consider the case that $\sup\{i : x(i)=1\}<\infty$ and $\inf\{i : y(i)=1\}>-\infty$. Clearly, $\psi_{\rm canc}(x,y)$ is well-defined for such x,y, even though both may be infinite.

and hence $y \in \mathcal{S}_{\text{fin}}(\Lambda)$ by the continuity of ϕ. Since we can change x outside $\mathcal{R}(\phi)$ without changing $\phi(x)$, and $\phi(e_i) = 0$ for $i \in \Lambda \backslash \mathcal{R}(\phi)$, we now have

$$\phi(x) = \phi\Big(\bigoplus_{\substack{i \in \mathcal{R}(\phi) \\ x(i)=1}} e_i \Big) = \bigoplus_{\substack{i \in \mathcal{R}(\phi) \\ x(i)=1}} \phi(e_i) = \bigoplus_{\substack{i \in \Lambda \\ x(i)=1}} \phi(e_i) = \psi_{\text{canc}}(x, y).$$

The rest of the proof is the same as the proof of Lemma 6.3. ∎

We now consider an interacting particle system whose generator G has a random mapping representation of the form (6.1) such that all local maps $m \in \mathcal{G}$ are cancellative. We also assume that the rates satisfy (4.11) so that the stochastic flow $(\mathbf{X}_{s,t})_{s \leq t}$ and the backward stochastic flow $(\mathbf{F}_{t,s})_{t \geq s}$ are well-defined. Since the concatenation of two cancellative functions is again cancellative, $(\mathbf{F}_{t,s})_{t \geq s}$ maps the space $\mathcal{C}_{\text{canc}}(S^\Lambda, \{0, 1\})$ into itself, so using Lemma 6.12, just as we did in the additive case (recall (6.15)), we can define a backward stochastic flow $(\mathbf{Y}_{t,s})_{t \geq s}$ on $\mathcal{S}_{\text{fin}}(\Lambda)$ by

$$\mathbf{F}_{t,s}(\psi_{\text{canc}}(\,\cdot\,, y)) =: \psi_{\text{canc}}\big(\,\cdot\,, \mathbf{Y}_{t,s}(y)\big) \qquad \big(t \geq s,\ y \in \mathcal{S}_{\text{fin}}(\Lambda)\big). \tag{6.27}$$

We will see that by the same arguments as in the additive case, $(\mathbf{Y}_{t,s})_{t \geq s}$ is the backward stochastic flow of a cancellative particle system $(Y_t)_{t \geq 0}$ that is pathwise dual to the system $(X_t)_{t \geq 0}$ with generator G. The following lemma, that will be proved below, is similar to Lemma 6.4.

Lemma 6.13 (Dual maps) *For each local cancellative map $m \colon \mathcal{S}(\Lambda) \to \mathcal{S}(\Lambda)$, there exists a unique map $\tilde{m} \colon \mathcal{S}(\Lambda) \to \mathcal{S}(\Lambda)$ that is dual to m with respect to the duality function ψ_{canc}, in the sense that*

$$\psi_{\text{canc}}\big(m(x), y\big) = \psi_{\text{canc}}\big(x, \tilde{m}(y)\big) \qquad \big(x, y \in \mathcal{S}(\Lambda),\ |x| \wedge |y| < \infty\big). \tag{6.28}$$

This dual map is also local and cancellative and uniquely characterized by

$$m(e_i)(j) = 1 \quad \Leftrightarrow \quad \tilde{m}(e_j)(i) = 1 \qquad (i, j \in \Lambda). \tag{6.29}$$

In terms of our graphical way of depicting cancellative maps, formula (6.17) can be described by saying that, just as in the additive case,

$\tilde{m}$ is obtained from m by keeping the blocking symbols and reversing the direction of all arrows. (6.30)

Using this, we see that the cancellative duals of the maps listed at the beginning of the section are given by:

$$\begin{gathered} \widetilde{\mathtt{abra}}_{ij} = \mathtt{abra}_{ji}, \quad \widetilde{\mathtt{death}}_i = \mathtt{death}_i, \\ \widetilde{\mathtt{vot}}_{ij} = \mathtt{arw}_{ji}, \quad \widetilde{\mathtt{arw}}_{ij} = \mathtt{vot}_{ji}, \quad \widetilde{\mathtt{excl}}_{ij} = \mathtt{excl}_{ij}. \end{gathered} \tag{6.31}$$

Note that the voter map is both additive and cancellative and has two different dual maps depending on whether we are considering additive or cancellative duality. The main result about cancellative duality is the following analogue of Theorem 6.5.

Theorem 6.14 (Cancellative duality) *Let G be the generator of an interacting particle system $(X_t)_{t \geq 0}$ with state space $\mathcal{S}(\Lambda)$. Assume that G has a random mapping representation* *of the form (6.1) such that all local maps $m \in \mathcal{G}$ are cancellative and the rates $(r_m)_{m\in\mathcal{G}}$ satisfy (4.11). Then*

$$\tilde{G} := \sum_{m\in\mathcal{G}} r_m\{f(\tilde{m}(y)) - f(y)\} \qquad (y \in \mathcal{S}_{\text{fin}}(\Lambda)) \tag{6.32}$$

is the generator of a nonexplosive continuous-time Markov chain $(Y_t)_{t\geq 0}$ with state space $\mathcal{S}_{\text{fin}}(\Lambda)$. Let ω be a graphical representation associated with the random mapping representation (6.1) of G. Define a graphical representation $\tilde{\omega}$ associated with the random mapping representation (6.32) of $\tilde{G}$ by

$$\tilde{\omega} := \{(\tilde{m}, t) : (m, t) \in \omega\}.$$

Let $(\mathbf{X}_{s,t})_{s\leq t}$ be the stochastic flow on $\mathcal{S}(\Lambda)$ defined in terms of ω as in Theorem 4.14, and let $(\mathbf{Y}_{t,s})_{t\geq s}$ be the backward stochastic flow on $\mathcal{S}_{\text{fin}}(\Lambda)$ defined in terms of $\tilde{\omega}$ as in Theorem 2.24. Then almost surely

$$\psi_{\text{canc}}(\mathbf{X}_{s,t}(x), y) = \psi_{\text{canc}}(x, \mathbf{Y}_{t,s}(y)) \qquad (s \leq t,\ x \in \mathcal{S}(\Lambda),\ y \in \mathcal{S}_{\text{fin}}(\Lambda)). \tag{6.33}$$

If the random mapping representation (6.21) also satisfies (4.11), then $(\mathbf{Y}_{t,s})_{t\geq s}$ can be extended to $\mathcal{S}(\Lambda)$ and $(\mathbf{X}_{s,t})_{s\leq t}$ maps the space $\mathcal{S}_{\text{fin}}(\Lambda)$ into itself. In this case, (6.33) also holds for $x \in \mathcal{S}_{\text{fin}}(\Lambda)$ and $y \in \mathcal{S}(\Lambda)$.

Proof of Lemma 6.13 As in the additive case, it is easy to see that a cancellative map is local if and only if it is defined by finitely many arrows and blocking symbols. Using the recipe "reverse the arrows, keep the blocking symbols", we can find a local map $\tilde{m}$ such that (6.29) holds. To see that it is dual to m in the sense of (6.28), let Δ be the set of lattice points where a blocking symbol is located or that are the starting point or endpoint of an arrow. Let x' be the restriction of x to Δ, that is, $x'(i) := x(i)$ if $i \in \Delta$ and $:= 0$ otherwise. Similarly, let y' denote the restriction of y to Δ, and let x'' and y'' denote the restrictions of x and y to $\Lambda\backslash\Delta$. Then

$$\psi_{\text{canc}}(m(x), y) = \psi_{\text{canc}}(m(x'), y') \oplus \psi_{\text{canc}}(x'', y''),$$

where

$$\psi_{\text{canc}}(m(x'), y') = \psi_{\text{canc}}\Big(\bigoplus_{i:\,x'(i)=1} m(e_i), y'\Big) = \bigoplus_{i:\,x'(i)=1}\ \bigoplus_{j:\,y'(j)=1} 1_{\{m(e_i)(j) = 1\}}.$$

Rewriting $\psi_{\text{canc}}(x, \tilde{m}(y))$ in the same way, using (6.29), we see that $\tilde{m}$ is dual to m in the sense of (6.28). As in the additive case, it is easy to see that $\tilde{m}$ is uniquely determined by (6.28). ∎

Proof of Theorem 6.14 This is completely the same as the proof of Theorem 6.5, except at the very end. If the random mapping representation (6.32) also satisfies (4.11), then by using what is already proved with the roles of the forward and backward processes reversed, we see that $(\mathbf{X}_{s,t})_{s\leq t}$ maps the space $\mathcal{S}_{\text{fin}}(\Lambda)$ into itself, and (6.33) also holds for $x \in \mathcal{S}_{\text{fin}}(\Lambda)$ and $y \in \mathcal{S}(\Lambda)$. ∎

The following lemma is similar to Lemma 6.6.

Lemma 6.15 (Distribution determining functions) *The class of functions $\{\psi_{\text{canc}}(\,\cdot\,, y) : y \in \mathcal{S}_{\text{fin}}(\Lambda)\}$ is distribution determining on $\mathcal{S}(\Lambda)$.*

Proof We may equivalently prove that the functions

$$g_y(x) := 1 - 2\psi_{\rm canc}(x, y) = (-1)^{\sum_i x(i)y(i)}$$

are distribution determining. Since $g_y g_{y'} = g_{y\oplus y'}$, the class $\{g_y : y \in \mathcal{S}_{\rm fin}(\Lambda)\}$ is closed under products, and since $g_{e_i}(x) = (-1)^{x(i)}$, this class separates points. The claim now follows from Lemma 4.39. ∎

Additive and cancellative duality are so similar that one wonders if they can be treated in a unified way. This is indeed the case. For local state spaces with three or more elements, an obvious thing one can do is to replace the addition modulo two from cancellative systems by addition modulo three or more. There are also less obvious possibilities. The paper [LS21] explores dualities where $(\{0,1\}, \vee)$ or $(\{0,1\}, \oplus)$ are replaced by commutative monoids or semirings.

Some models that a priori do not look like cancellative systems turn out to be representable in cancellative maps. An example is the Neuhauser–Pacala model, defined by its transition rates in (1.18). We define a *rebellious map* by

$$\texttt{rebel}_{ijk}(x)(l) := \begin{cases} x(i) \oplus x(j) \oplus x(k) & \text{if } l = k, \\ x(l) & \text{otherwise.} \end{cases} \tag{6.34}$$

In words, this says that $x(k)$ changes its state if $x(i) \neq x(j)$.

Exercise 6.16 *Show that the map* $\texttt{rebel}_{ijk}$ *is cancellative. Show that the generator of the Neuhauser–Pacala model defined in (1.18) can be represented as*

$$\begin{aligned} G_{\rm NP} f(x) &= \frac{\alpha}{|\mathcal{N}_i|} \sum_i \sum_{j \in \mathcal{N}_i} \{f(\texttt{vot}_{ji}(x)) - f(x)\} \\ &= \frac{1-\alpha}{|\mathcal{N}_i|^2} \sum_i \sum_{k,j \in \mathcal{N}_i} \{f(\texttt{rebel}_{kji}(x)) - f(x)\}. \end{aligned}$$

Exercise 6.17 *In the* threshold voter model, *the site i changes its type $x(i)$ from 0 to 1 with rate one as long as at least one site in its neighborhood $\mathcal{N}_i$ has type 1, and likewise, i flips from 1 to 0 with rate one as long as at least one site in $\mathcal{N}_i$ has type 0. Show that the generator of the threshold voter model can be written as*

$$G_{\rm thres} f(x) = 2^{-|\mathcal{N}_i|+1} \sum_i \sum_{\substack{\Delta \subset \mathcal{N}_i \cup \{i\} \\ |\Delta| \text{ is even}}} \{f(m_{\Delta,i}(x)) - f(x)\},$$

where $m_{\Delta,i}$ is the cancellative map defined by

$$m_{\Delta,i}(x)(k) := \begin{cases} x(i) \oplus \bigoplus_{j\in\Delta} x(j) & \text{if } k = i, \\ x(k) & \text{otherwise.} \end{cases}$$

Cancellative duality for the threshold voter model is used extensively in [Han99].

Exercise 6.18 *Show that the threshold voter model is monotone.*

6.5 Lloyd–Sudbury Duality

The additive systems duality function (6.13) and cancellative systems duality function (6.25) are not the only choices of ψ that lead to useful dualities. There are two approaches to finding useful duality functions: the pathwise approach, that aims to find dualities between stochastic flows in the sense of (6.5), and the algebraic approach, that only aims to prove distributional relations of the form (6.7). There has been a lot of recent work on the algebraic approach, starting with the paper [GK+09], linking dualities to representations of Lie algebras. For an overview of this work, we refer to [GR25]. We will present some older results below, based on the algebraic approach, due to Lloyd and Sudbury [SL95, SL97, Sud99].

The pathwise approach always depends on finding a clever stochastic flow and then finding a suitable space of functions on S^Λ that is mapped into itself by the stochastic flow of the backward in time process. As we have seen, for additive and cancellative systems, the spaces $\mathcal{C}_{\rm add}(\{0,1\}^\Lambda,\{0,1\})$ and $\mathcal{C}_{\rm canc}(\{0,1\}^\Lambda,\{0,1\})$ are invariant, and this naturally leads to additive and cancellative duality.

To explain a bit about the algebraic approach, which only aims to prove relations of the form (6.7) without proving a duality of stochastic flows, for technical simplicity, for the remainder of this section, we will restrict ourselves to finite state spaces. In general, when trying to prove a duality for interacting particle systems on infinite lattices, it is often a good idea to first prove the result on finite lattices and then extend it to infinite lattices using approximation results such as Theorem 4.43 and Corollary 4.44. We will demonstrate this method in Section 6.6.

As we have done before, we write $\mathbb{E}^x$ (respectively $\mathbb{E}^y$) to denote expectation with respect to the law of the process X started in $X_0 = x$ (respectively started in $Y_0 = y$).

Lemma 6.19 (Duality of finite Markov processes) *Let $(X_t)_{t\geq 0}$ and $(Y_t)_{t\geq 0}$ be Markov processes with finite state spaces S and R, generators G and H, and Markov semigroups $(P_t)_{t\geq 0}$ and $(Q_t)_{t\geq 0}$. Then one has*

$$\mathbb{E}^x\big[\psi(X_t,y)\big] = \mathbb{E}^y\big[\psi(x,Y_t)\big] \qquad (x\in S,\ y\in R,\ t\geq 0) \tag{6.35}$$

if and only if

$$G\psi(\,\cdot\,,y)(x) = H\psi(x,\,\cdot\,)(y) \qquad (x\in S,\ y\in R). \tag{6.36}$$

Proof The duality relation (6.35) says that

$$\sum_{x'\in S} P_t(x,x')\psi(x',y) = \sum_{y'\in R}\psi(x,y')Q_t(y,y') \qquad (x\in S,\ y\in R,\ t\geq 0),$$

which can in matrix form be written as

$$P_t\psi = \psi Q_t^\dagger \qquad (t\geq 0), \tag{6.37}$$

where $Q_t^\dagger(y',y) := Q_t(y,y')$ denotes the transpose of Q_t. Differentiating with respect to t and setting $t=0$, it follows that

$$G\psi = \psi H^\dagger$$

which is just a more formal way of writing (6.36). Conversely, if (6.36) holds, then $G^2\psi =$

$G\psi H^\dagger = \psi(H^\dagger)^2$ and by induction $G^n\psi = \psi(H^\dagger)^n$ for all $n \geq 0$. Using the fact that

$$P_t = \sum_{n=0}^{\infty} \frac{1}{n!} t^n G^n \quad \text{and} \quad Q_t = \sum_{n=0}^{\infty} \frac{1}{n!} t^n H^n,$$

it follows that $P_t\psi = \psi Q_t^\dagger$ $(t \geq 0)$, and hence (6.35) holds. ∎

Let $(X_t)_{t\geq 0}$ and $(Y_t)_{t\geq 0}$ be Markov processes with finite state spaces S and R, generators G and H, and Markov semigroups $(P_t)_{t\geq 0}$ and $(Q_t)_{t\geq 0}$. Let K be a probability kernel from S to R. A relation of the form (compare (6.37))

$$P_t K = K Q_t \qquad (t \geq 0) \tag{6.38}$$

is called an *intertwining* of Markov processes. Note that (6.38) says that the following two procedures are equivalent for each S-valued random variable X_0:

- Evolve the state X_0 for time t under the evolution of the Markov process $(X_t)_{t\geq 0}$, then map the outcome X_t into a random variable Y_t using the kernel K.
- Map X_0 into a random variable Y_0 using the kernel K, then evolve Y_0 for time t under the evolution of the Markov process $(Y_t)_{t\geq 0}$.

We can summarize the situation in the following commutative diagram:

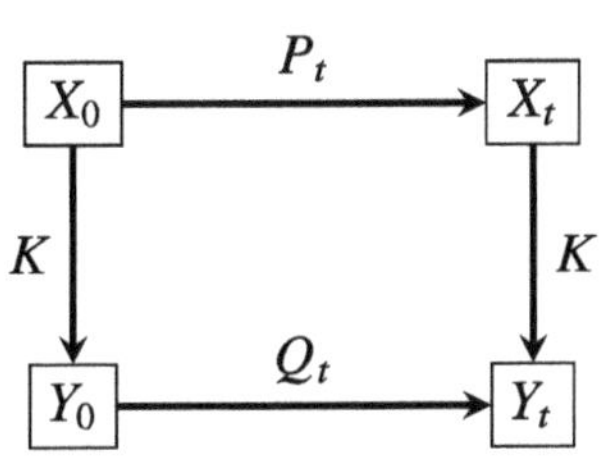

(6.39)

Lemma 6.20 (Intertwining of finite Markov processes) *The intertwining relation (6.38) is equivalent to*

$$GK = KH. \tag{6.40}$$

Proof Analogue to the proof of Lemma 6.19. ∎

As one might guess, there is a close relationship between duality and intertwining. If $(P_t)_{t\geq 0}$, $(Q_t)_{t\geq 0}$, and $(R_t)_{t\geq 0}$ are Markov semigroups, K is a probability kernel, and ψ a duality function such that

$$P_t K = K Q_t \quad \text{and} \quad Q_t\psi = \psi R_t^\dagger \qquad (t \geq 0),$$

then trivially

$$P_t(K\psi) = KQ_t\psi = (K\psi)R_t^\dagger \qquad (t \geq 0), \tag{6.41}$$

which says that the Markov processes with semigroups $(P_t)_{t\geq 0}$ and $(R_t)_{t\geq 0}$ are dual with duality function $K\psi$.

To see these general principles at work, let us look at interacting particle systems with state space of the form $\{0,1\}^\Lambda$, where Λ is finite. For each $r > 0$, we let ψ_r denote the duality function

$$\psi_r(x,y) := \prod_{i\in\Lambda}(1-r)^{x(i)y(i)} \qquad \big(x,y \in \{0,1\}^\Lambda\big). \tag{6.42}$$

Using the fact that $0^n = 1_{\{n=0\}}$, we observe that

$$\left.\begin{aligned} \psi_1(x,y) &= 1-\psi_{\text{add}}(x,y) \\ \psi_2(x,y) &= (-1)^{\psi_{\text{canc}}(x,y)} \end{aligned}\right\} \qquad \big(x,y \in \{0,1\}^\Lambda\big).$$

Therefore, duality with respect to duality functions of the form (6.42) includes additive and cancellative duality as special cases. Duality functions of the form (6.42) arose from the work of Lloyd and Sudbury [SL95, SL97]. Accordingly, we will call ψ_r the *Lloyd–Sudbury duality function.*

There is a close connection between duality functions of the form (6.42) and *thinning*, as we now explain. Thinning has already been introduced in Section 2.5, but for convenience we repeat the definition here. Let $(\chi_p(i))_{i\in\Lambda}$ be i.i.d. with $\mathbb{P}[\chi_p(i) = 1] = p$ and $\mathbb{P}[\chi_p(i) = 0] = 1-p$. Then

$$K_p(x,y) := \mathbb{P}\big[y(i) = \chi_p(i)x(i)\ \forall i \in \Lambda\big] \qquad \big(x,y \in \{0,1\}^\Lambda\big) \tag{6.43}$$

defines a *thinning kernel.* Note that if we interpret sites i with $x(i) = 1$ as being occupied by a particle, then the effect of K_p is to independently throw away some of these particles, where each particle has a probability p to remain. We claim that

$$K_pK_q = K_{pq} \quad \text{and} \quad K_p\psi_r = \psi_{pr} \qquad (0 \le p,q \le 1,\ r > 0). \tag{6.44}$$

The first relation is clear from the interpretation in terms of thinning, while the second relation follows by writing

$$\begin{aligned} K_p\psi_r(x,z) &= \sum_y K_p(x,y)\prod_{i\in\Lambda}(1-r)^{y(i)z(i)} = \mathbb{E}\Big[\prod_{i\in\Lambda}(1-r)^{\chi_p(i)x(i)z(i)}\Big] \\ &= \prod_{i\in\Lambda}\mathbb{E}\big[(1-r)^{\chi_p(i)x(i)z(i)}\big] = \prod_{i\in\Lambda}(1-pr)^{x(i)z(i)} = \psi_{pr}(x,z). \end{aligned}$$

The following lemma says that if a particle system has two duals, one with respect to the duality function ψ_{r_1} and the other with respect to the duality function ψ_{r_2}, then one of these duals is a thinning of the other.

Lemma 6.21 (Lloyd–Sudbury duals and thinning) *Let G, H_1, and H_2 be generators of Markov processes with state space $\{0,1\}^\Lambda$ where Λ is finite. Let $0 < r_1 \le r_2$ and set $p := r_1/r_2$. Then of the relations*

$$\text{(i) } H_1\psi_{r_1} = \psi_{r_1}G^\dagger, \quad \text{(ii) } H_2\psi_{r_2} = \psi_{r_2}G^\dagger, \quad \textit{and} \quad \text{(iii) } H_1K_p = K_pH_2$$

any two imply the third.

Proof Using all three relations (i)–(iii) as well as (6.44), we have the "circular" sequence of equalities:

$$H_1\psi_{r_1} \overset{\text{(i)}}{=} \psi_{r_1}G^\dagger = K_p\psi_{r_2}G^\dagger \overset{\text{(ii)}}{=} K_pH_2\psi_{r_2} \overset{\text{(iii)}}{=} H_1K_p\psi_{r_2} = H_1\psi_{r_1}.$$

From this, we immediately see that of the relations

$$\text{(i) } H_1\psi_{r_1} = \psi_{r_1}G^{\dagger}, \quad \text{(ii)}' \ K_pH_2\psi_{r_2} = K_p\psi_{r_2}G^{\dagger}, \quad \text{(iii)}' \ H_1K_p\psi_{r_2} = K_pH_2\psi_{r_2}$$

any two imply the third. To complete the proof, it suffices to show that K_p and ψ_r are invertible as matrices for all $p \in (0,1]$ and $r > 0$, since we can then multiply (ii)′ from the left with K_p^{-1} and (iii)′ from the right with $\psi_{r_2}^{-1}$ to obtain (ii) and (iii).

We can view the linear space of all functions $f\colon \{0,1\}^{\Lambda} \to \mathbb{R}$ as the tensor product $\bigotimes_{i\in\Lambda}\mathbb{R}^{\{0,1\}}$. In this picture, the matrices K_p and ψ_r are the tensor product over Λ of single-site matrices of the form

$$\begin{pmatrix} 1 & 0 \\ 1-p & p \end{pmatrix} \quad \text{and} \quad \begin{pmatrix} 1 & 1 \\ 1 & 1-r \end{pmatrix},$$

respectively. These single-site matrices are invertible for all $p \in (0,1]$ and $r \in (0,\infty)$ and hence the same is true for their tensor products K_p and ψ_r. ∎

It is useful to look at a concrete example. Let (Λ, E) be a finite graph, as in (1.2) let $\mathcal{N}_i := \big\{j \in \Lambda : \{i,j\} \in E\big\}$ denote the neighborhood of a vertex $i \in \Lambda$, and assume that $N := |\mathcal{N}_i|$ does not depend on $i \in \Lambda$. In line with notation introduced in Section 1.1, let $\mathcal{E} := \big\{(i,j) \in \Lambda^2 : \{i,j\} \in E\big\}$ denote the set of directed edges associated with E. Let $G_{\rm vot}, G_{\rm rw}\ G_{\rm arw}$ be the Markov generators defined by

$$\begin{aligned} G_{\rm vot}f(x) &:= N^{-1}\sum_{(i,j)\in\mathcal{E}} \big\{f\big(\mathtt{vot}_{ij}(x)\big) - f\big(x\big)\big\}, \\ G_{\rm rw}f(x) &:= N^{-1}\sum_{(i,j)\in\mathcal{E}} \big\{f\big(\mathtt{rw}_{ij}(x)\big) - f\big(x\big)\big\}, \\ G_{\rm arw}f(x) &:= N^{-1}\sum_{(i,j)\in\mathcal{E}} \big\{f\big(\mathtt{arw}_{ij}(x)\big) - f\big(x\big)\big\}, \end{aligned}$$

where the voter model map $\mathtt{vot}_{ij}$, the coalescing random walk map $\mathtt{rw}_{ij}$, and the annihilating random walk map $\mathtt{arw}_{ij}$ are defined in (1.4), (1.24), and (1.26), respectively. In words, $G_{\rm vot}$ is the generator of a voter model in which each site $i \in \Lambda$ adopts with rate one the type of a randomly chosen neighbor. The processes with generators $G_{\rm rw}$ and $G_{\rm arw}$ consist of coalescing and annihilating particles that jump to a randomly chosen neighboring site with rate one. We claim that

$$\text{(i) } G_{\rm rw}\psi_1 = \psi_1G^{\dagger}_{\rm vot}, \quad \text{(ii) } G_{\rm arw}\psi_2 = \psi_2G^{\dagger}_{\rm vot}, \quad \text{(iii) } G_{\rm rw}K_{1/2} = K_{1/2}G_{\rm arw}.$$

Indeed, the voter model map is both additive and cancellative, so (i) follows from Theorem 6.5 since the coalescing random walk map is the additive dual of the voter model map and likewise (ii) follows from Theorem 6.14 since the annihilating random walk map is the cancellative dual of the voter model map. By Lemma 6.21, (i) and (ii) imply (iii), which says that annihilating random walks are a 1/2-thinning of coalescing random walks. In other words, for each $t \geq 0$, the following two procedures are equivalent:

- Run coalescing random walk dynamics for time t and then thin the resulting configuration with 1/2.
- Thin the initial configuration with 1/2 and then run annihilating random walk dynamics for time t.

One can also verify this directly and use this to deduce (ii) from (i) or vice versa. In Proposition 6.24, we will see a less trivial example of a thinning relation between two interacting particle systems.

Surprisingly, there exist many duality relations between interacting particle systems with respect to the Lloyd–Sudbury duality function ψ_r for other values of r than $r = 1, 2$. These dualities can usually not be obtained as pathwise dualities.

Let (Λ, E) be a finite graph. The paper [Sud00] considers interacting particle systems on graphs where the configuration along each edge makes the following transitions with the following rates:[4]

$$\begin{array}{rccll} \text{“annihilation”} & & 11 \mapsto 00 & & \text{at rate } a, \\ \text{“branching”} & 01 \mapsto 11 & \text{and} & 10 \mapsto 11 & \text{each at rate } b, \\ \text{“coalescence”} & 11 \mapsto 01 & \text{and} & 11 \mapsto 10 & \text{each at rate } c, \\ \text{“death”} & 01 \mapsto 00 & \text{and} & 10 \mapsto 00 & \text{each at rate } d, \\ \text{“exclusion”} & 01 \mapsto 10 & \text{and} & 10 \mapsto 01 & \text{each at rate } e. \end{array}$$

More formally, for each $i, j \in \Lambda$, we can define a map $m_{ij}^{01\mapsto 11}$ on $\{0, 1\}^\Lambda$ as follows:

$$m_{ij}^{01\mapsto 11}(x)(k) = \begin{cases} 1 & \text{if } k = i \text{ and } \big(x(i), x(j)\big) = (0, 1), \\ 1 & \text{if } k = j \text{ and } \big(x(i), x(j)\big) = (0, 1), \\ x(k) & \text{in all other cases.} \end{cases}$$

Defining $m_{ij}^{11\mapsto 00}$, and so forth, in a similar way, the generator of the process we are interested in can be written as

$$\begin{aligned} Gf(x) = &\sum_{\{i,j\}} a\big\{f\big(m_{ij}^{11\mapsto 00}(x)\big) - f(x)\big\} \\ &+ \sum_{(i,j)} \Big[b\big\{f\big(m_{ij}^{01\mapsto 11}(x)\big) - f(x)\big\} + c\big\{f\big(m_{ij}^{11\mapsto 01}(x)\big) - f(x)\big\} \\ &\qquad + d\big\{f\big(m_{ij}^{01\mapsto 00}(x)\big) - f(x)\big\} + e\big\{f\big(m_{ij}^{01\mapsto 10}(x)\big) - f(x)\big\}\Big], \end{aligned} \tag{6.45}$$

where the first sum runs over all (unordered) edges $\{i, j\} \in E$ and the second sum runs over all ordered pairs (i, j) such that $\{i, j\} \in E$.

Theorem 6.22 (Lloyd–Sudbury duality) *Let G and G' be defined as in (6.45) in terms of rates a, b, c, d, e and a', b', c', d', e', respectively, and let $r > 0$. Then one has*

$$G\psi_r = \psi_r G'^{\dagger} \tag{6.46}$$

if and only if $a' = a + 2(1 - r)\gamma$, $b' = b + \gamma$, $c' = c - (2 - r)\gamma$, $d' = d + \gamma$, and $e' = e - \gamma$, where $\gamma := (a + c - d + (1 - r)b)/r$.

Proof This follows from Lemma 6.19 by checking (6.36). The calculations are a bit tedious,

[4] The meaning of the words “annihilation”, “branching”,. . . here is a bit different from the way we have used these words so far. In particular, the “death” rate d refers only to “deaths while the neighboring site is empty”, while “deaths while the neighboring site is occupied” are called “coalescence”.

so we omit them here. They can be found in [Sud00, formula (9)], which is a simplification of [SL95, formula (21)]. ∎

We note that a generalization of Theorem 6.22 to directed graphs can be found in [Swa06, Prop. 10].

6.6 The Contact-Voter Model

As we have already seen, $\psi_1 = 1 - \psi_{\rm add}$ and $\psi_2 = (-1)^{\psi_{\rm canc}}$ correspond to additive and cancellative duality. It seems that for $r \neq 1, 2$, dualities of the form (6.46) are almost never[5] pathwise dualities. To give an example with $r \neq 1, 2$, consider an interacting particle system on a (possibly infinite) graph (Λ, E) whose dynamics are a mixture of contact process and voter model dynamics, with generator of the form:

$$\begin{aligned} G_{\rm covo} f(x) :=\, & \lambda \sum_{(i,j)\in\mathcal{E}} \big\{f\big(\mathtt{bra}_{ij}(x)\big) - f(x)\big\} + \sum_{i\in\Lambda} \big\{f\big(\mathtt{death}_i(x)\big) - f(x)\big\} \\ & + \alpha \sum_{(i,j)\in\mathcal{E}} \big\{f\big(\mathtt{vot}_{ij}(x)\big) - f(x)\big\} \qquad \big(x \in \{0,1\}^\Lambda\big), \end{aligned} \tag{6.47}$$

where $\mathcal{E}$ denotes the set of directed edges associated with E. Letting $\mathcal{N}_i := \big\{j \in \Lambda : \{i, j\} \in E\big\}$ denote the set of neighbors of i, we assume that Λ is countable and

$$\sup_{i\in\Lambda} |\mathcal{N}_i| < \infty,$$

which implies that the generator in (6.47) satisfies the summability condition (4.11) of Theorem 4.14 and hence corresponds to a well-defined interacting particle system. Such systems are studied in [DLZ14], who are especially interested in the fast-voting limit $\alpha \to \infty$. The contact-voter model is additive (but not cancellative, because the branching map is not), and by Theorem 6.5, dual with respect to the duality function $\psi_1 = 1 - \psi_{\rm add}$ to the interacting particle system with generator

$$\begin{aligned} G_{\rm corw} f(y) :=\, & \lambda \sum_{(i,j)\in\mathcal{E}} \big\{f\big(\mathtt{bra}_{ij}(y)\big) - f(y)\big\} + \sum_{i\in\Lambda} \big\{f\big(\mathtt{death}_i(y)\big) - f(y)\big\} \\ & + \alpha \sum_{(i,j)\in\mathcal{E}} \big\{f\big(\mathtt{rw}_{ij}(y)\big) - f(y)\big\} \qquad \big(y \in \{0,1\}^\Lambda\big), \end{aligned} \tag{6.48}$$

which corresponds to a system of branching and coalescing random walks. Perhaps surprisingly, the contact-voter model is also self-dual.

Proposition 6.23 (Self-duality of the contact-voter model) *Assume that $\lambda > 0$. Then the contact-voter model with generator as in (6.47) is self-dual with respect to the duality function ψ_r with $r := \lambda/(\alpha + \lambda)$.*

Proof We first consider the case that the graph (Λ, E) is finite. The generator $G_{\rm covo}$ is a special case of the generators considered in Theorem 6.22 and corresponds to the choice of parameters

$$a = 0, \quad b = \lambda + \alpha, \quad c = 1, \quad d = 1 + \alpha, \quad e = 0.$$

[5] Except some very trivial and pathological cases.

We observe that setting $r := \lambda/(\alpha + \lambda)$ makes the parameter γ from Theorem 6.22 zero, which has the effect that $a' = a$, $b' = b$, $c' = c$, $d' = d$, and $e' = e$; that is, we have found a self-duality.

To extend the result to infinite graphs, we use an approximation argument. We need to show that

$$\mathbb{E}\big[\psi_r(\mathbf{X}_{0,t}(x), x')\big] = \mathbb{E}\big[\psi_r(x, \mathbf{X}_{0,t}(x'))\big] \qquad (t \geq 0,\ x, x' \in \{0,1\}^\Lambda), \tag{6.49}$$

where $(\mathbf{X}_{s,u})_{s\leq u}$ denotes the stochastic flow defined by the graphical representation of the contact-voter model, and

$$\psi_r(x, y) := \prod_{i\in\Lambda}(1-r)^{x(i)y(i)} \qquad (x, y \in \{0,1\}^\Lambda). \tag{6.50}$$

The argument will be a bit tricky since it is in general not true that $x_n \to x$ and $y_n \to y$ pointwise imply that $\psi_r(x_n, y_n) \to \psi_r(x, y)$. This prevents us from using general approximation results like Corollary 4.44. We observe, however, that $x_n \uparrow x$ and $y_n \uparrow x$ imply $\psi_r(x_n, y_n) \downarrow \psi_r(x, y)$. This is why we will base our argument on monotone approximation.

Our first aim is to prove (6.49) for $x, x' \in \mathcal{S}_{\rm fin}(\Lambda)$, the set of finite configurations. Let (Λ_n, E_n) be finite subgraphs of (Λ, E) that increase to the whole graph. Let $(\mathbf{X}^n_{s,t})_{s\leq t}$ be the stochastic flow of a restricted process defined by a graphical representation in which we have removed all death maps outside Λ_n and all branching and voter maps along edges that are not in E_n. For all n large enough so that x and x' are zero outside Λ_n, we have that also $\mathbf{X}^n_{0,t}(x)$ and $\mathbf{X}^n_{0,t}(x')$ are zero outside Λ_n for all $t \geq 0$. Note that if x and y are zero outside Λ_n, then in (6.50) it does not matter if we take the product over Λ or Λ_n. Therefore, applying Theorem 6.22 to the processes on the finite graphs (Λ_n, E_n), we see that

$$\mathbb{E}\big[\psi_r(\mathbf{X}^n_{0,t}(x), x')\big] = \mathbb{E}\big[\psi_r(x, \mathbf{X}^n_{0,t}(x'))\big] \qquad (t \geq 0) \tag{6.51}$$

for all n large enough. By Theorem 2.35, the contact-voter model restricted to $\mathcal{S}_{\rm fin}(\Lambda)$ is a nonexplosive continuous-time Markov chain. It follows that almost surely, there exists a (random) $m < \infty$ such that for all $n \geq m$, the unrestricted process $(\mathbf{X}_{0,s}(x))_{0\leq s\leq t}$ stays inside Λ_n up to time t. But then $(\mathbf{X}^n_{0,s}(x))_{0\leq s\leq t}$ must be equal to $(\mathbf{X}_{0,s}(x))_{0\leq s\leq t}$ for all $n \geq m$ so we see that almost surely $\mathbf{X}^n_{0,t}(x) \to \mathbf{X}_{0,t}(x)$ as $n \to \infty$ with respect to the discrete topology on $\mathcal{S}_{\rm fin}(\Lambda)$, and by the same argument also $\mathbf{X}^n_{0,t}(x') \to \mathbf{X}_{0,t}(x')$ with respect to the discrete topology on $\mathcal{S}_{\rm fin}(\Lambda)$. Taking the limit $n \to \infty$ in (6.51), it follows that (6.49) holds for all $x, x' \in \mathcal{S}_{\rm fin}(\Lambda)$.

For general $x, x' \in \{0,1\}^\Lambda$, we can find $x_n, x'_n \in \mathcal{S}_{\rm fin}(\Lambda)$ such that $x_n \uparrow x$ and $x'_n \uparrow x'$. Then also $\mathbf{X}_{0,t}(x_n) \uparrow \mathbf{X}_{0,t}(x)$ and $\mathbf{X}_{0,t}(x'_n) \uparrow \mathbf{X}_{0,t}(x')$ by the monotonicity and continuity of $\mathbf{X}_{0,t}$. Using the continuity of ψ_r with respect to increasing sequences, we obtain (6.49) in full generality. ∎

We have already seen in Lemma 6.21 that there is a close connection between the Llyod–Sudbury duality functions ψ_r and thinning. The following proposition demonstrates this on our example of the contact-voter model.

Proposition 6.24 (Thinning of the contact-voter model) *Let $(P_t)_{t\geq 0}$ and $(Q_t)_{t\geq 0}$ denote the semigroups of the contact-voter model with generator as in (6.47) and the system of* *branching and coalescing random walks with generator as in (6.48), respectively. Let K_r denote the thinning kernel defined in (6.43) with $p := \lambda/(\alpha+\lambda)$. Then*

$$P_t K_p = K_p Q_t \qquad (t \geq 0). \tag{6.52}$$

Proof We first prove the statement for finite graphs. Additive duality tells us that (i) $G_{\text{covo}}\psi_1 = \psi_1 G^{\dagger}_{\text{corw}}$, and Proposition 6.23 tells us that (ii) $G_{\text{covo}}\psi_p = \psi_p G^{\dagger}_{\text{covo}}$. By Lemma 6.21, this implies (iii) $G_{\text{covo}}K_p = K_p G_{\text{corw}}$, which implies (6.52).

To also get the result for infinite graphs, we use approximation with finite graphs. In this case, the argument is simpler than in the proof of Proposition 6.23. We claim that thinning kernels are continuous, that is, $f \in \mathcal{C}(\{0,1\}^\Lambda)$ implies $K_r f \in \mathcal{C}(\{0,1\}^\Lambda)$. Indeed, if $f : \{0,1\}^\Lambda \to \mathbb{R}$ is continuous, $x_n \to x$ pointwise, and $(\chi_i(i))_{i\in\Lambda}$ are i.i.d. Bernoulli random variables with intensity r, then

$$K_r f(x_n) = \mathbb{E}\big[f(\chi_r x_n)\big] \underset{n\to\infty}{\longrightarrow} \mathbb{E}\big[f(\chi_r x)\big] = K_r f(x),$$

where $(\chi_r x)(i) := \chi_r(i)x(i)$ denotes the pointwise product of χ_r and x. It follows that if X^n are random variables with values in $\{0,1\}^\Lambda$ that converge weakly in law to X, and Y^n and Y are obtained from X^n and X by thinning with the kernel K_r, then the Y_n converge weakly in law to Y. As a result, we can use Corollary 4.44 to approximate infinite systems with finite systems and take the limit to get the result for infinite systems. ∎

We will continue our study of the contact-voter model in Section 6.9.

6.7 Invariant Laws of the Voter Model

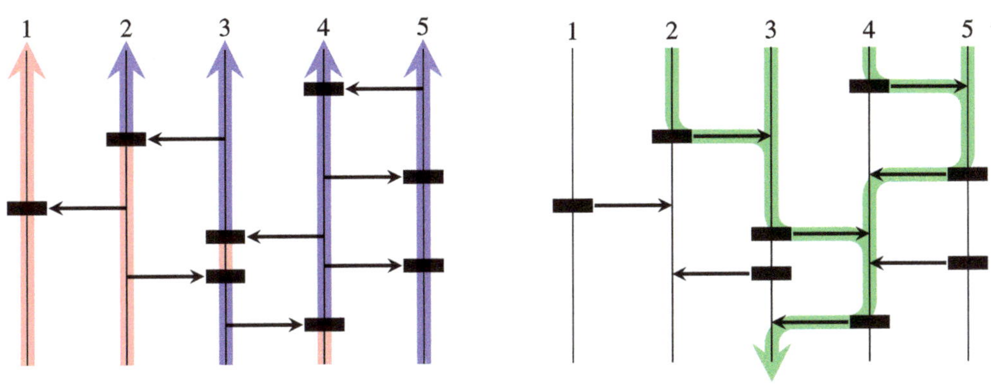

Figure 6.7 Graphical representation of a one-dimensional voter model and its dual system of coalescing random walks. At the final time, the points 2, 3, 4, and 5 have the same type because they descend from the same ancestor.

By Theorem 6.5, the voter model X is pathwise dual, with respect to the additive duality function ψ_{add} from (6.13), to a collection Y of coalescing random walks. Due to the fact that $|Y_t|$ is a nonincreasing function of t (i.e., the number of walkers can only decrease), it is much easier to work with this dual system than with the voter model itself, so duality is really the key to understanding the voter model.

Proposition 6.25 (Clustering in low dimensions) *Let X be a nearest-neighbor or range R voter model on $\mathbb{Z}^d$. Assume that $d = 1, 2$. Then, regardless of the initial law,*

$$\mathbb{P}[X_t(i) = X_t(j)] \underset{t\to\infty}{\longrightarrow} 1 \qquad \forall i, j \in \mathbb{Z}^d.$$

Moreover, the delta measures $\delta_{\underline{0}}$ and $\delta_{\underline{1}}$ on the constant configurations are the only extremal invariant laws.

Proof In the graphical representation of the voter model, for each $(i, t) \in \mathbb{Z}^d \times \mathbb{R}$ and $s \geq 0$, there is a unique site

$$j =: \xi_s^{(i,t)} \in \mathbb{Z}^d \text{ such that } (j, t - s) \rightsquigarrow (i, t).$$

Here $(\xi_s^{(i,t)})_{s\geq 0}$ is the path of a random walk starting at $\xi_0^{(i,t)} = i$ and "running downwards in the graphical representation". Two such random walks started from different space-time points (i, t) and (i', t') are independent up to the first time they meet and coalesce as soon as they meet. Moreover, if $X_t = \mathbf{X}_{0,t}(X_0)$, then, as demonstrated in Figure 6.7,

$$X_t(i) = X_{t-s}(\xi_s^{(i,t)}) \qquad (0 \leq s \leq t),$$

that is, $\xi_s^{(i,t)}$ traces back where the site i at time t got its type from.[6]

Since the difference $\xi_s^{(i,t)} - \xi_s^{(j,t)}$ of two such random walks is a random walk with absorption in the origin, and since random walk on $\mathbb{Z}^d$ in dimensions $d = 1, 2$ is recurrent, we observe that

$$\mathbb{P}[X_t(i) = X_t(j)] \geq \mathbb{P}[\xi_t^{(i,t)} = \xi_t^{(j,t)}] = \mathbb{P}[\xi_t^{(i,0)} = \xi_t^{(j,0)}] \underset{t\to\infty}{\longrightarrow} 1 \qquad \forall i, j \in \mathbb{Z}^d.$$

This clearly implies that all invariant laws must be concentrated on constant configurations; that is, a general invariant law is of the form $p\delta_{\underline{0}} + (1 - p)\delta_{\underline{1}}$ with $p \in [0, 1]$. ∎

For product initial laws, we can be more precise. Although we state the following theorem for two-type processes only, it is clear from the proof that the statement generalizes basically unchanged to multitype voter models.

Theorem 6.26 (Process started in product law) *Let X be a nearest neighbor or range R voter model on $\mathbb{Z}^d$. Assume that the $(X_0(i))_{i\in\mathbb{Z}^d}$ are i.i.d. with intensity $\mathbb{P}[X_0(i) = 1] = p \in [0, 1]$. Then*

$$\mathbb{P}[X_t \in \cdot\,] \underset{t\to\infty}{\Longrightarrow} \nu_p, \tag{6.53}$$

where ν_p is an invariant law of the process. If $d = 1, 2$, then

$$\nu_p = (1 - p)\delta_{\underline{0}} + p\delta_{\underline{1}}. \tag{6.54}$$

On the other hand, if $d \geq 3$ and $0 < p < 1$, then the measures ν_p are concentrated on configurations that are not constant.

[6] This construction works in fact generally for multitype voter models, where the local state space S can be any finite set, and which are in general of course not additive systems. For simplicity, we will focus on the two-type voter model here.

Proof As in the proof of Proposition 6.25, let $(\xi_s^{(i,t)})_{s\geq 0}$ be the backward random walk in the graphical representation starting at (i,t). Define a random equivalence relation $\sim$ on $\mathbb{Z}^d$ by

$$i \sim j \quad \text{if and only if} \quad \xi_s^{(i,0)} = \xi_s^{(j,0)} \text{ for some } s \geq 0.$$

We claim that if we color the equivalence classes of $\sim$ in an i.i.d. fashion such that each class gets the color 1 with probability p and the color 0 with probability $1-p$, then this defines an invariant law ν_p such that (6.53) holds. Since random walk in dimensions $d = 1, 2$ is recurrent, there is a.s. only one equivalence class, and $\nu_p = (1-p)\delta_{\underline{0}} + p\delta_{\underline{1}}$. On the other hand, since random walk in dimensions $d \geq 3$ is transient, there are a.s. infinitely many[7] equivalence classes, and hence for $p \neq 0, 1$ the measure ν_p is concentrated on configurations that are not constant.

To prove (6.53), we use coupling. Let $(\chi(i))_{i\in\mathbb{Z}^d}$ be i.i.d. $\{0,1\}$-valued with $\mathbb{P}[\chi(i) = 1] = p$. For each $t \geq 0$, we define a random equivalence relation $\sim_t$ on $\mathbb{Z}^d$ by

$$i \sim_t j \quad \text{if and only if} \quad \xi_s^{(i,0)} = \xi_s^{(j,0)} \text{ for some } 0 \leq s \leq t.$$

We enumerate the elements of $\mathbb{Z}^d$ in some arbitrary way and define

$$\tilde{X}_t(i) := \chi(j) \quad \text{where } j \text{ is the smallest element of } \{k \in \mathbb{Z}^d : i \sim_t k\}. \tag{6.55}$$

Then $\tilde{X}_t$ is equally distributed with X_t and converges a.s. as $t \to \infty$ to a random variable with law ν_p. ∎

Remark In dimensions $d \geq 3$, it is in fact known that the measures ν_p are extremal, and each extremal invariant law of the voter model is of this form. See [Lig85, Thm V.1.8].

Exercise 6.27 *Let $(Y_t)_{t\geq 0}$ be coalescing random walks with generator as in (1.25). Show that the upper invariant law is $\delta_{\underline{0}}$, the delta-measure on the all-zero configuration.* Hint: *Use Lemma 4.33 to derive a differential equation for $\mathbb{E}^{\overline{\nu}}[Y_t(0)]$. To complete the argument, you will need to argue that if $\overline{\nu} \neq \delta_{\underline{0}}$, then the event $\{y : y(0) = y(1) = 1\}$ has positive probability under $\overline{\nu}$.*

Exercise 6.28 *Prove that the voter model started in a finite initial state dies out:*

$$\mathbb{P}^x\big[\exists t \geq 0 \text{ s.t. } X_t = \underline{0}\big] = 1 \qquad \big(x \in \mathcal{S}_{\text{fin}}(\mathbb{Z}^d)\big).$$

Hint: *You can use the previous exercise and duality. Alternatively, you can use martingale convergence.*

Note that the statements of both previous exercises are not true on finite lattices.

[7] Although this is intuitively plausible, it requires a bit of work to prove this. A quick proof, that however requires a bit of ergodic theory, is as follows: Since Poisson point processes are spatially ergodic, and the number N of equivalence classes is a translation-invariant random variable, this random number N must in fact be a.s. constant. Since the probability that two paths coalesce tends to zero as the distance between their starting points tends to infinity, for each finite n, we can find n starting points sufficiently far from each other so that with positive probability, none of the paths started at these points coalesce. This implies that $\mathbb{P}[N \geq n] > 0$ for each finite n, and hence, by the fact that N is a.s. constant, $\mathbb{P}[N = \infty] = 1$.

6.8 Homogeneous Invariant Laws

In the present section, we show how the self-duality of the contact process can be used to prove that for contact processes with some sort of translation invariant structure, the upper invariant law is the limit law started from any nontrivial translation-invariant initial law, and we will show that this in turn implies that the function $\theta(\lambda)$ from (5.8) is continuous everywhere, except possibly at the critical point. The methods of the present section are not restricted to additive particle systems. Applications of the technique to cancellative systems can be found in [BDD91, SS08, CP14], while [LS23a] treats a coupling of an additive and a cancellative system.

We work in the setup of Section 5.5, so we consider contact processes with generators of the form (5.7), where Λ is a countable set, p is a symmetric probability kernel on Λ such that $p(i,i) = 0$ $(i \in \Lambda)$, and the pair (Λ, p) is vertex transitive as defined in (5.6). We will also assume that p is irreducible in the sense that for each $i, j \in \Lambda$, there exists an $n \geq 0$ such that $p^n(i, j) > 0$. We start with a simple observation that has been anticipated before, and that says that the functions $\theta(\lambda)$ from (1.9) and (5.8) are the same. We continue to use the notation $|x| := \sum_i x(i)$ and let $\mathcal{S}_{\text{fin}}(\Lambda) := \{x \in \{0,1\}^\Lambda : |x| < \infty\}$ denote the space of finite configurations.

Lemma 6.29 (The function theta) *Let X denote the contact process with generator of the form (5.7), and let $\overline{\nu}$ denote its upper invariant law. Then*

$$\int \overline{\nu}(\mathrm{d}x)\, x(i) = \mathbb{P}^{e_i}[X_t \neq \underline{0}\ \forall t \geq 0] \qquad (i \in \Lambda).$$

More generally, for any $y \in \mathcal{S}_{\text{fin}}(\Lambda)$,

$$\int \overline{\nu}(\mathrm{d}x)\, 1_{\{x \wedge y \neq \underline{0}\}} = \mathbb{P}^y[X_t \neq \underline{0}\ \forall t \geq 0].$$

Proof By Theorem 6.5, the contact process X is self-dual with respect to the additive systems duality function, that is,

$$\mathbb{P}^x[X_t \wedge y = \underline{0}] = \mathbb{P}^y[x \wedge X_t = \underline{0}] \qquad \big(t \geq 0,\ x, y \in \{0,1\}^\Lambda\big).$$

In particular, setting $x = \underline{1}$, we see that for any $y \in \mathcal{S}_{\text{fin}}(\Lambda)$,

$$\int \overline{\nu}(\mathrm{d}x)\, 1_{\{x \wedge y \neq \underline{0}\}} = \lim_{t\to\infty} \mathbb{P}^{\underline{1}}[X_t \wedge y \neq \underline{0}] = \lim_{t\to\infty} \mathbb{P}^y[\underline{1} \wedge X_t \neq \underline{0}] = \mathbb{P}^y[X_t \neq \underline{0}\ \forall t \geq 0].$$

Note that since $|y| < \infty$, the function $x \mapsto 1_{\{x \wedge y \neq \underline{0}\}}$ is continuous, which together with the weak convergence $\delta_{\underline{1}} P_t \Rightarrow \overline{\nu}$ implies the first equality above. The condition $|y| < \infty$ can be removed by doing this step more carefully, using monotone convergence instead of weak convergence, but since contact processes started in infinite initial states a.s. do not die out in finite time, this case is less interesting. ∎

We will be interested in processes that are started in a translation-invariant initial law. Since we are working in a rather general setup, we have to say more precisely what we mean by translation invariance. Let $\mathrm{Aut}(\Lambda, p)$ be the group of all automorphisms of (Λ, p), in the sense defined in Section 5.5. We say that a subgroup $\mathcal{T} \subset \mathrm{Aut}(\Lambda, p)$ is *vertex transitive* if for each $i, j \in \Lambda$, there exists a $\psi \in \mathcal{T}$ such that $\psi(i) = j$. To see an example, consider the case that $\Lambda = \mathbb{Z}^d$ and p is the transition kernel of a symmetric nearest-neighbor random walk on $\mathbb{Z}^d$. For each $j \in \mathbb{Z}^d$, let ψ_j be the translation defined as $\psi_j(i) := i + j$ ($i \in \mathbb{Z}^d$). Then the group of translations $\mathcal{T} := \{\psi_j : j \in \mathbb{Z}^d\}$ is vertex transitive. In this example, $\mathcal{T}$ is smaller than the group $\mathrm{Aut}(\mathbb{Z}^d, p)$ of all automorphisms of $(\mathbb{Z}^d, p)$, which also contains rotations, inversions, and more.

From now on, we fix a vertex transitive subgroup $\mathcal{T} \subset \mathrm{Aut}(\Lambda, p)$. We say that a probability law μ on $\{0,1\}^\Lambda$ is *homogeneous* or *translation invariant* if $\mu \circ \psi^{-1} = \mu$ for all $\psi \in \mathcal{T}$. The main aim of the present section is to prove the following result, which is originally due to Harris [Har76], with a similar result for a one-dimensional discrete-time process already having been proved by Vasil'ev [Vas69]. We can think of this result as a sort of spatial analogue of the observation in Section 3.5 that for the mean-field contact process, solutions of the differential equation (3.23) started in any nonzero initial state converge to the upper fixed point. Recall from Section 5.5 that a probability law μ on $\{0,1\}^\Lambda$ is *nontrivial* if $\mu(\{\underline{0}\}) = 0$, that is, if μ gives zero probability to the all-zero configuration.

Theorem 6.30 (Convergence to upper invariant law) *Let $(X_t)_{t\geq 0}$ be a contact process started in a homogeneous nontrivial initial law $\mathbb{P}[X_0 \in \cdot\,]$. Then*

$$\mathbb{P}[X_t \in \cdot\,] \underset{t\to\infty}{\Longrightarrow} \overline{\nu},$$

where $\overline{\nu}$ is the upper invariant law.

We start with two preparatory lemmas. We will use the graphical representation of the contact process as an additive particle system (see Section 6.2) and use the shorthand

$$X_t^x := \mathbf{X}_{0,t}(x) \qquad \big(t \geq 0,\ x \in \{0,1\}^\Lambda\big),$$

where $(\mathbf{X}_{s,t})_{s\leq t}$ is the stochastic flow constructed from the graphical representation.

Lemma 6.31 (Extinction versus unbounded growth) *For each $x \in \mathcal{S}_{\mathrm{fin}}(\Lambda)$, one has*

$$X_t^x = \underline{0} \text{ for some } t \geq 0 \quad \text{or} \quad |X_t^x| \underset{t\to\infty}{\longrightarrow} \infty \quad \text{a.s.} \tag{6.56}$$

Proof Define

$$\rho(x) := \mathbb{P}\big[X_t^x \neq \underline{0}\ \forall t \geq 0\big] \qquad \big(x \in \mathcal{S}_{\mathrm{fin}}(\Lambda)\big).$$

Since there is a positive probability that each infected site dies before it manages to reproduce, it is not hard to see that for each $N \geq 0$, there exists an $\varepsilon > 0$ such that

$$|x| \leq N \quad \text{implies} \quad \rho(x) \leq 1 - \varepsilon. \tag{6.57}$$

We first argue why it is plausible that this implies (6.56) and then give a rigorous proof. Imagine that $|X_t^x| \not\to \infty$. Then, in view of (6.57), the process infinitely often gets a chance of at least ε to die out; hence, eventually it should die out.

To make this rigorous, let

$$\mathcal{A}_x := \{X_t^x \neq \underline{0}\ \forall t \geq 0\} \qquad \big(x \in \mathcal{S}_{\mathrm{fin}}(\Lambda)\big)$$

denote the event that the process $(X_t^x)_{t\geq 0}$ survives, and let $\mathcal{F}_t$ be the σ-field generated by the Poisson point processes used in our graphical representation till time t. Then

$$\rho(X_t^x) = \mathbb{P}\big[\mathcal{A}_x \,\big|\, \mathcal{F}_t\big] \underset{t\to\infty}{\longrightarrow} 1_{\mathcal{A}_x} \quad \text{a.s.,} \tag{6.58}$$

where we have used an elementary result from probability theory that says that if $\mathcal{F}_n$ is an increasing sequence of σ-fields and $\mathcal{F}_\infty = \sigma(\bigcup_n \mathcal{F}_n)$, then $\lim_{n\to\infty} \mathbb{P}[\mathcal{A}|\mathcal{F}_n] = \mathbb{P}[\mathcal{A}|\mathcal{F}_\infty]$ a.s. for each measurable event $\mathcal{A}$. (See [Loe63, § 29, Complement 10 (b)].) In view of (6.57), formula (6.58) implies (6.56). ∎

Lemma 6.32 (Nonzero intersection) *Let $(X_t)_{t\geq 0}$ be a contact process with a homogeneous nontrivial initial law $\mathbb{P}[X_0 \in \cdot\,]$. Then for each $s, \varepsilon > 0$, there exists an $N \geq 1$ such that for any $y \in \mathcal{S}_{\rm fin}(\Lambda)$*

$$|y| \geq N \quad \textit{implies} \quad \mathbb{P}\big[X_s \wedge y = \underline{0}\big] \leq \varepsilon.$$

Proof We construct $(X_t)_{t\geq 0}$ as $X_t := \mathbf{X}_{0,t}(X_0)$ $(t \geq 0)$, where X_0 is independent of the graphical representation ω. We let $(\mathbf{Y}_{t,s})_{t\geq s}$ denote the backward stochastic flow of the dual process as in Theorem 6.5. For each $\delta > 0$, we let p_δ denote the matrix defined by $p_\delta(i,j) := 1_{\{p(i,j)\geq\delta\}}p(i,j)$ $(i, j \in \Lambda)$, and for $m \geq 1$ we set

$$\Lambda_m(i) := \big\{j \in \Lambda : p^n_{1/m}(i,j) > 0 \text{ for some } 0 \leq n \leq m\big\}.$$

We fix an arbitrary reference point $0 \in \Lambda$. By vertex transitivity, $|\Lambda_m(i)| = |\Lambda_m(0)|$ does not depend on $i \in \Lambda$. It is not hard to see that for each $y \in \{0,1\}^\Lambda$ with $|y| \geq N$ we can find a $y' \leq y$ with $|y'| \geq N/|\Lambda_m(0)|$ such that the sets $\Lambda_m(i)$ where i ranges through $\{i : y'(i) = 1\}$ are disjoint. We let $(\mathbf{Y}^{m,i}_{t,s})_{t\geq s}$ denote the backward stochastic flow of the dual process restricted to $\Lambda_m(i)$. More precisely, this is the stochastic flow associated with the modified graphical representation obtained by removing all branchings from inside $\Lambda_m(i)$ to its complement. Then, using Hölder's inequality[8] in the inequality marked with an exclamation mark, we have

$$\begin{aligned}
\mathbb{P}\big[X_s \wedge y = \underline{0}\big] &= \mathbb{P}\big[X_0 \wedge \mathbf{Y}_{s,0}(y) = \underline{0}\big] = \int \mathbb{P}[X_0 \in \mathrm{d}x]\,\mathbb{P}\big[x \wedge \mathbf{Y}_{s,0}(y) = \underline{0}\big] \\
&\leq \int \mathbb{P}[X_0 \in \mathrm{d}x]\,\mathbb{P}\Big[x \wedge \bigvee_{i:\, y'(i)=1} \mathbf{Y}^{m,i}_{s,0}(e_i) = \underline{0}\Big] \\
&= \int \mathbb{P}[X_0 \in \mathrm{d}x] \prod_{i:\, y'(i)=1} \mathbb{P}\big[x \wedge \mathbf{Y}^{m,i}_{s,0}(e_i) = \underline{0}\big] \\
&\overset{!}{\leq} \prod_{i:\, y'(i)=1} \Big(\int \mathbb{P}[X_0 \in \mathrm{d}x]\,\mathbb{P}\big[x \wedge \mathbf{Y}^{m,i}_{s,0}(e_i) = \underline{0}\big]^{|y'|}\Big)^{1/|y'|} \\
&= \prod_{i:\, y'(i)=1} \Big(\int \mathbb{P}[X_0 \in \mathrm{d}x]\,\mathbb{P}\big[x \wedge \mathbf{Y}^{m,0}_{s,0}(e_0) = \underline{0}\big]^{|y'|}\Big)^{1/|y'|} \\
&= \int \mathbb{P}[X_0 \in \mathrm{d}x]\,\mathbb{P}\big[x \wedge \mathbf{Y}^{m,0}_{s,0}(e_0) = \underline{0}\big]^{|y'|},
\end{aligned}$$

where we have used the homogeneity of $\mathbb{P}[X_0 \in \cdot\,]$ in the penultimate equality. Our arguments so far show that $|y| \geq N$ implies that

$$\mathbb{P}\big[X_s \wedge y = \underline{0}\big] \leq \int \mathbb{P}[X_0 \in \mathrm{d}x]\mathbb{P}\big[x \wedge \mathbf{Y}^{m,0}_{s,0}(e_0) = \underline{0}\big]^{N/|\Lambda_m(0)|} =: f(N,m).$$

Here, using the fact that

$$\mathbb{P}\big[x \wedge \mathbf{Y}^{m,0}_{s,0}(e_0) = \underline{0}\big] < 1 \quad \text{if } x(i) = 1 \text{ for some } i \in \Lambda_m(0),$$

[8] Recall that Hölder's inequality says that $1/p + 1/q = 1$ implies $\|fg\|_1 \leq \|f\|_p\|g\|_q$, where $\|f\|_p := (\int |f|^p \mathrm{d}\mu)^{1/p}$. By induction, this gives $\|\prod_{i=1}^n f_i\|_1 \leq \prod_{i=1}^n \|f_i\|_n$.

we see that

$$\lim_{N\uparrow\infty} f(N,m) = \int \mathbb{P}[X_0 \in \mathrm{d}x] 1_{\{x(i)=0\ \forall i\in\Lambda_m(0)\}} = \mathbb{P}[X_0(i) = 0\ \forall i \in \Lambda_m(0)].$$

Since the kernel p is irreducible, we have that $\Lambda_m(0) \uparrow \Lambda$, so by the nontriviality of $\mathbb{P}[X_0 \in \cdot\,]$ it follows that

$$\lim_{m\uparrow\infty} \mathbb{P}[X_0(i) = 0\ \forall i \in \Lambda_m(0)] = \mathbb{P}[X_0 = \underline{0}] = 0.$$

Together with our previous equation, this shows that

$$\lim_{m\to\infty}\lim_{N\to\infty} f(N,m) = 0.$$

For each $\varepsilon > 0$, we can first choose m large enough such that $\lim_{N\to\infty} f(N,m) \le \varepsilon/2$ and then N large enough such that $f(N,m) \le \varepsilon$, proving our claim. ∎

Exercise 6.33 *Show by counterexample that the statement of Lemma 6.32 is false for $s = 0$.*

Proof of Theorem 6.30 As in the proof of Lemma 6.31, we set

$$\rho(x) := \mathbb{P}\big[X^x_t \neq \underline{0}\ \forall t \ge 0\big] \qquad \big(x \in \mathcal{S}_{\rm fin}(\Lambda)\big).$$

We construct $(X_t)_{t\ge0}$ as $X_t := \mathbf{X}_{0,t}(X_0)$ $(t \ge 0)$, where X_0 is independent of the graphical representation. By Lemmas 4.40, 6.6, and 6.29, it suffices to show that

$$\lim_{t\to\infty} \mathbb{P}\big[\mathbf{X}_{0,t}(X_0) \wedge y \neq \underline{0}\big] = \rho(y) \qquad \big(y \in \mathcal{S}_{\rm fin}(\Lambda)\big).$$

By duality, for any $s > 0$, this is equivalent to

$$\lim_{t\to\infty} \mathbb{P}\big[\mathbf{X}_{0,s}(X_0) \wedge \mathbf{Y}_{t,s}(y) \neq \underline{0}\big] = \rho(y) \qquad \big(y \in \mathcal{S}_{\rm fin}(\Lambda)\big).$$

Setting $X_s := \mathbf{X}_{0,s}(X_0)$ and $X^y_t := \mathbf{Y}_{s+t,s}(y)$, we may equivalently show that

$$\lim_{t\to\infty} \mathbb{P}\big[X_s \wedge X^y_t \neq \underline{0}\big] = \rho(y) \qquad \big(y \in \mathcal{S}_{\rm fin}(\Lambda)\big),$$

where X_s and X^y_t are independent, and $s > 0$ is some fixed constant. For each $\varepsilon > 0$, we can choose N as in Lemma 6.32 and write

$$\begin{aligned}\mathbb{P}\big[X_s \wedge X^y_t \neq \underline{0}\big] =\,& \mathbb{P}\big[X_s \wedge X^y_t \neq \underline{0}\,\big|\,|X^y_t| = 0\big]\,\mathbb{P}\big[|X^y_t| = 0\big]\\ &+\mathbb{P}\big[X_s \wedge X^y_t \neq \underline{0}\,\big|\,0 < |X^y_t| < N\big]\,\mathbb{P}\big[0 < |X^y_t| < N\big]\\ &+\mathbb{P}\big[X_s \wedge X^y_t \neq \underline{0}\,\big|\,|X^y_t| \ge N\big]\,\mathbb{P}\big[|X^y_t| \ge N\big].\end{aligned}$$

Here, by Lemma 6.31 and our choice of N,

$$\begin{aligned}&\text{(i)} && \mathbb{P}\big[X_s \wedge X^y_t \neq \underline{0}\,\big|\,|X^y_t| = 0\big] = 0,\\ &\text{(ii)} && \lim_{t\to\infty}\mathbb{P}\big[0 < |X^y_t| < N\big] = 0,\\ &\text{(iii)} && \liminf_{t\to\infty}\mathbb{P}\big[X_s \wedge X^y_t \neq \underline{0}\,\big|\,|X^y_t| \ge N\big] \ge 1-\varepsilon,\\ &\text{(iv)} && \lim_{t\to\infty}\mathbb{P}\big[|X^y_t| \ge N\big] = \rho(x),\end{aligned}$$

from which we conclude that

$$(1-\varepsilon)\rho(x) \le \liminf_{t\to\infty}\mathbb{P}\big[X_s \wedge X^y_t \neq \underline{0}\big] \le \limsup_{t\to\infty}\mathbb{P}\big[X_s \wedge X^y_t \neq \underline{0}\big] \le \rho(x).$$

Since $\varepsilon > 0$ is arbitrary, our proof is complete. ∎

Theorem 6.30 has a simple corollary.

Corollary 6.34 (Homogeneous invariant laws) *All homogeneous invariant laws of a contact process are convex combinations of $\delta_{\underline{0}}$ and $\overline{\nu}$.*

Proof Let ν be any homogeneous invariant law. We will show that ν is a convex combination of $\delta_{\underline{0}}$ and $\overline{\nu}$. If $\nu = \delta_{\underline{0}}$, we are done. Otherwise, as in the proof of Lemma 5.19, we can write $\nu = (1-p)\delta_{\underline{0}} + p\mu$, where $p \in (0,1]$ and μ is a nontrivial homogeneous invariant law. But now Theorem 6.30 implies that

$$\mu = \mu P_t \underset{t\to\infty}{\Longrightarrow} \overline{\nu},$$

so we conclude that $\mu = \overline{\nu}$. ∎

Recall from Exercise 5.23 that the function $\lambda \mapsto \theta(\lambda)$ from (5.8) is right-continuous everywhere. We let

$$\lambda_{\mathrm{c}} := \inf\{\lambda \in \mathbb{R} : \theta(\lambda) > 0\} \tag{6.59}$$

denote the *critical point* of the contact process. As an application of Theorem 6.30, we prove the following result.

Proposition 6.35 (Continuity above the critical point) *The function $\lambda \mapsto \theta(\lambda)$ is left-continuous on $(\lambda_{\mathrm{c}}, \infty)$.*

Proof Let $\overline{\nu}_\lambda$ denote the upper invariant law of the contact process with infection rate λ. Fix $\lambda > \lambda_{\mathrm{c}}$ and choose $\lambda_{\mathrm{c}} < \lambda_n \uparrow \lambda$. Since the space $\mathcal{M}_1(\{0,1\}^\Lambda)$ of probability measures on $\{0,1\}^\Lambda$, equipped with the topology of weak convergence, is compact, it suffices to show that each subsequential limit ν_* of the measures $\overline{\nu}_{\lambda_n}$ equals $\overline{\nu}_\lambda$. By Proposition 4.45, each subsequential limit ν_* is an invariant law. It is clearly also homogeneous. Since $\lambda > \lambda_{\mathrm{c}}$, by Lemma 5.19, the measures $\overline{\nu}_{\lambda_n}$ are nontrivial for all n, and hence, using also Proposition 5.20, the same is true for ν_*. By Corollary 6.34, we conclude that $\nu_* = \overline{\nu}$. This argument shows that the map

$$(\lambda_{\mathrm{c}}, \infty) \ni \lambda \mapsto \overline{\nu}_\lambda$$

is left-continuous w.r.t. the topology of weak convergence. Since $x \mapsto x(i)$ is a continuous function and $\theta(\lambda)$ is its expectation under $\overline{\nu}_\lambda$, the claim follows. ∎

Exercise 6.36 *Let $(X_t)_{t\geq 0}$ be an additive interacting particle system, and let $(Y_t)_{t\geq 0}$ be its additive dual. Show that the upper invariant law $\overline{\nu}$ of $(X_t)_{t\geq 0}$ is uniquely characterized by*

$$\int \overline{\nu}(\mathrm{d}x)\, 1_{\{x \wedge y \neq \underline{0}\}} = \mathbb{P}^y[Y_t \neq \underline{0}\ \forall t \geq 0] \qquad \big(y \in \mathcal{S}_{\mathrm{fin}}(\Lambda)\big). \tag{6.60}$$

Exercise 6.37 *Let $(X_t)_{t\geq 0}$ be a cancellative interacting particle system, and let $(Y_t)_{t\geq 0}$ be it cancellative dual. Let $\pi_{1/2}$ denote product measure with intensity $1/2$. Show that*

$$\mathbb{P}^{\pi_{1/2}}\big[X_t \in \cdot\,\big] \underset{t\to\infty}{\Longrightarrow} \nu_{1/2},$$

where $\nu_{1/2}$ is an invariant law that is uniquely characterized by the relation

$$\int \nu_{1/2}(\mathrm{d}x)1_{\{|x \wedge y| \text{ is odd}\}} = \tfrac{1}{2}\mathbb{P}^y\big[Y_t \neq \underline{0}\ \forall t \geq 0\big] \qquad \big(y \in \mathcal{S}_{\mathrm{fin}}(\Lambda)\big).$$

Because of the similarity of this formula to the characterization of the upper invariant law of an additive interacting particle system in (6.60), the measure $\nu_{1/2}$ is sometimes called the odd upper invariant law.

Exercise 6.38 *If we drop the assumption that the probability kernel in (5.7) is symmetric, then the contact process is no longer self-dual. Show that in such a setting, vertex transitivity implies that the constant*

$$K := \sum_{i \in \Lambda} p(i, j) \tag{6.61}$$

does not depend on $j \in \Lambda$. Assuming that $K < \infty$, show that the dual process is a contact process with kernel $p'(i, j) := K^{-1}p(j, i)$ and infection rate $\lambda' := K\lambda$. Give an example of a transitive pair (Λ, p) for which $K \neq 1$. Hint: *Consider an infinite tree in which each vertex has three neighbors. Give each edge an orientation so that at each vertex, there are two incoming edges and one outgoing edge, and the oriented paths starting at any two vertices eventually meet. (Compare Exercise 6.9.)*

Exercise 6.39 *Show that Theorem 6.30 remains true if we drop the assumption that p is symmetric but assume that the constant in (6.61) is finite.*

6.9 Equality of Critical Points

The contact-voter model X on $\mathbb{Z}^d$, that has a mixture of contact process and voter model dynamics, has been introduced in Section 6.6. It has two parameters: the infection rate λ and the voter rate α. We say that X *survives* if

$$\mathbb{P}^{e_0}[X_t \neq \underline{0}\ \forall t \geq 0] > 0.$$

For each $\alpha \geq 0$, we define critical infection rates $\lambda_c(\alpha)$ and $\lambda'_c(\alpha)$ by

$$\begin{aligned}\lambda_c(\alpha) &:= \inf\big\{\lambda \in \mathbb{R} : \text{the upper invariant law is nontrivial}\big\},\\ \lambda'_c(\alpha) &:= \inf\big\{\lambda \in \mathbb{R} : \text{the process survives}\big\}.\end{aligned}$$

The paper [DLZ14] studies the asymptotics of $\lambda_c(\alpha)$ as $\alpha \to \infty$. Here, we will use duality to prove a simpler statement, namely, that $\lambda_c(\alpha) = \lambda'_c(\alpha)$ for all $\alpha \geq 0$.

For $\alpha = 0$ (that is, the pure contact process), we already know this, as it is a direct consequence of Lemma 6.29, which follows from the self-duality of the contact process. We will use a similar argument here using Proposition 6.23, which says that the contact-voter model is self-dual with respect to the duality function ψ_r from (6.42) with $r := \lambda/(\alpha + \lambda)$. Note that if $\alpha = 0$ (the pure contact process), then $r = 1$, which corresponds to additive systems duality.

Proposition 6.40 (Characterization of the upper invariant law) *Let $r := \lambda/(\alpha + \lambda)$. The* *upper invariant law* $\overline{\nu}$ *of the contact-voter model satisfies*

$$\int \overline{\nu}(\mathrm{d}x)\psi_r(x,y) = \mathbb{P}^y\big[X_t = \underline{0} \textit{ for some } t \geq 0\big] \tag{6.62}$$

for all finite $y \in \{0,1\}^{\mathbb{Z}^d}$. *Moreover, the upper invariant law is nontrivial if and only if the process survives. As a consequence,* $\lambda_c(\alpha) = \lambda'_c(\alpha)$ *for all* $\alpha \geq 0$.

Proof Letting $X^{\underline{1}}$ and X^y denote the processes started in $X^{\underline{1}}_0 = \underline{1}$ and $X^y_0 = y$, we observe that by Proposition 6.23,

$$\int \overline{\nu}(\mathrm{d}x)\,\psi_r(x,y) = \lim_{t\to\infty} \mathbb{E}\big[\psi_r(X^{\underline{1}}_t, y)\big] = \lim_{t\to\infty} \mathbb{E}\big[\psi_r(\underline{1}, X^y_t)\big] = \lim_{t\to\infty} \mathbb{E}\big[(1-r)^{|X^y_t|}\big].$$

The proof of Lemma 6.31 carries over without a change to the contact-voter model, so

$$X^y_t = \underline{0} \text{ for some } t \geq 0 \quad \text{or} \quad |X^y_t| \underset{t\to\infty}{\longrightarrow} \infty \quad \text{a.s.}$$

Using this, we see that

$$\lim_{t\to\infty} \mathbb{E}\big[(1-r)^{|X^y_t|}\big] = \mathbb{P}\big[X^y_t = \underline{0} \text{ for some } t \geq 0\big],$$

completing the proof of (6.62).

Inserting $y = e_0$ into (6.62), we see that

$$\int \overline{\nu}(\mathrm{d}x)\,(1-r)^{x(0)} = \mathbb{P}^{e_0}\big[X_t = \underline{0} \text{ for some } t \geq 0\big],$$

or equivalently, using the fact that $1 - (1-r)^{x(0)} = rx(0)$ with $r = \lambda/(\alpha+\lambda)$,

$$\frac{\lambda}{\alpha+\lambda}\int \overline{\nu}(\mathrm{d}x)\,x(0) = \mathbb{P}^{e_0}\big[X_t \neq \underline{0}\ \forall t \geq 0\big].$$

This shows that $\overline{\nu} \neq \delta_{\underline{0}}$ if and only if the process survives. ∎

7

Oriented Percolation

7.1 Introduction

Although we have seen phase transitions in our simulations of interacting particle systems in Chapter 1, and we have seen how phase transitions are defined and can be calculated in the mean-field limit in Chapter 3, we have not yet proved the existence of a phase transition for any of the spatial models that we have seen so far.

In the present chapter, we fill this gap by proving that the contact process on $\mathbb{Z}^d$ with generator as in (1.8) and death rate $\delta = 1$ undergoes a phase transition. We will show that the critical point λ_c defined in (6.59) is nontrivial in the sense that $0 < \lambda_c < \infty$. Note that by Lemma 6.29,

$$\begin{aligned}\lambda_c &= \inf\{\lambda \in \mathbb{R} : \text{ the contact process survives}\}\\ &= \inf\{\lambda \in \mathbb{R} : \text{ the upper invariant law is nontrivial}\}.\end{aligned}$$

In Exercise 5.21, which is based on Theorem 4.36, we have already proved that[1]

$$\frac{1}{|\mathcal{N}_0|} \le \lambda_c,$$

where $|\mathcal{N}_0| = 2d$ or $= (2R + 1)^d - 1$ is the size of the neighborhood of the origin for the nearest-neighbor process and for the range R process, respectively. In view of this, it suffices to prove that $\lambda_c < \infty$. A simple comparison argument (Exercise 5.27) shows that if the nearest-neighbor one-dimensional contact process survives for some value of λ, then the same is true for the nearest-neighbor and range R processes in dimensions $d \ge 2$. Thus, it suffices to show that $\lambda_c < \infty$ for the nearest-neighbor process in dimension one.

The method we will use is comparison with oriented percolation. This neither leads to a particularly short proof nor does it yield a very good upper bound on λ_c, but it has the advantage that it is a very robust method that can be applied to many other interacting particle systems. For example, in [SS08] and [SS15], the method is applied to rebellious voter models and systems with cooperative branching and coalescing random walk dynamics, respectively. An important paper for propagating the technique was [Dur91], where this was for the first time applied to non-monotone systems, and it was shown that "basically, all one needs" to prove survival is that a particle system spreads into empty areas at a positive speed.

[1] The apparent difference between this formula and the formula in Exercise 5.21 is due to the different conventions in (1.8) and (5.7).

7.2 Oriented Percolation

In order to prepare for the proof that the critical infection rate of the contact process is finite, in the present section, we will study *oriented* (or *directed*) *bond percolation* on $\mathbb{Z}^d$. For $i, j \in \mathbb{Z}^d$, we write $i \leq j$ if $i = (i_1, \dots, i_d)$ and $j = (j_1, \dots, j_d)$ satisfy $i_k \leq j_k$ for all $k = 1, \dots, d$. Let

$$\mathcal{A} := \{(i, j) : i, j \in \mathbb{Z}^d,\ i \leq j,\ |i - j| = 1\}. \tag{7.1}$$

We view $\mathbb{Z}^d$ as an infinite directed graph, where elements $(i, j) \in \mathcal{A}$ represent arrows (or *directed bonds*) between neighboring sites. Note that all arrows point "upwards" in the sense of the natural order on $\mathbb{Z}^d$. See Figure 7.1.

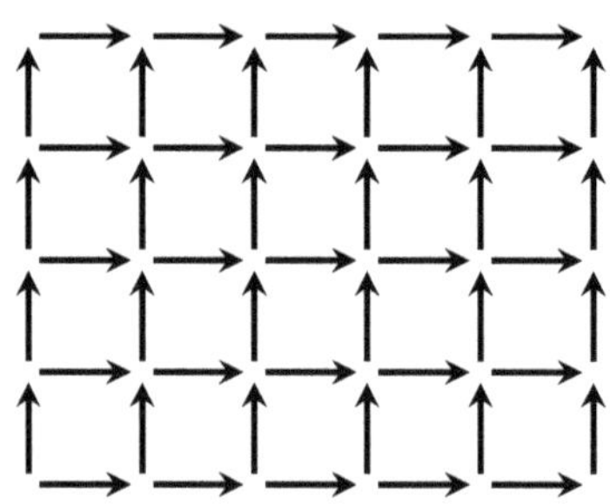

Figure 7.1 $\mathbb{Z}^2$ as an oriented graph.

Now fix some *percolation parameter* $p \in [0, 1]$ and let $(\omega_{(i,j)})_{(i,j)\in\mathcal{A}}$ be a collection of i.i.d. Bernoulli random variables with $\mathbb{P}[\omega_{(i,j)} = 1] = p$. We say that there is an *open path* from a site $i \in \mathbb{Z}^d$ to $j \in \mathbb{Z}^d$ if there exist $n \geq 0$ and a function $\gamma\colon \{0, \dots, n\} \to \mathbb{Z}^d$ such that $\gamma(0) = i$, $\gamma(n) = j$, and

$$(\gamma(k-1), \gamma(k)) \in \mathcal{A} \quad \text{and} \quad \omega_{(\gamma(k-1),\gamma(k))} = 1 \qquad (k = 1, \dots, n).$$

We denote the presence of an open path by $\rightsquigarrow$. Note that open paths must walk upwards in the sense of the order on $\mathbb{Z}^d$. We write $0 \rightsquigarrow \infty$ to indicate the existence of an infinite open path starting at the origin $0 \in \mathbb{Z}^d$. See Figure 7.2.

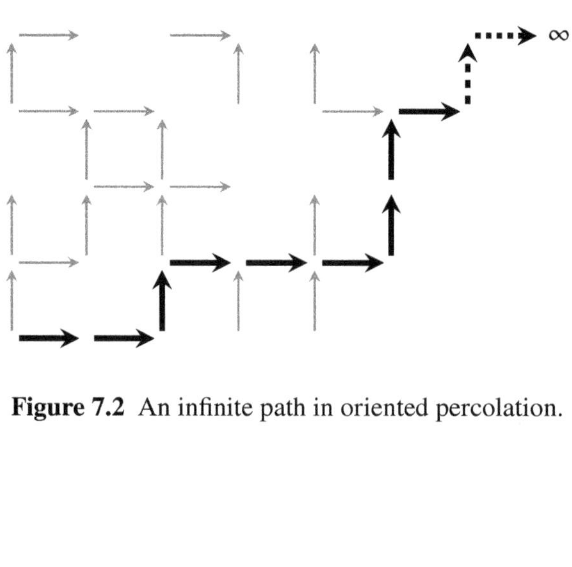

Figure 7.2 An infinite path in oriented percolation.

Exercise 7.1 *Show that the number of vertices that can be reached by an open path from the origin is infinite if and only if there starts an infinite open path at the origin.*

Theorem 7.2 (Critical percolation parameter) *For oriented percolation in dimensions $d \geq 2$, there exists a critical parameter $p_c = p_c(d)$ such that $\mathbb{P}[0 \rightsquigarrow \infty] = 0$ for $p < p_c$ and $\mathbb{P}[0 \rightsquigarrow \infty] > 0$ for $p > p_c$. One has*

$$\frac{1}{d} \leq p_c(d) \leq \frac{8}{9}.$$

Proof Set

$$p_c := \inf\{p \in [0,1] : \mathbb{P}[0 \rightsquigarrow \infty] > 0\}.$$

A simple monotone coupling argument shows that $\mathbb{P}[0 \rightsquigarrow \infty] = 0$ for $p < p_c$ and $\mathbb{P}[0 \rightsquigarrow \infty] > 0$ for $p > p_c$.

To prove that $0 < p_c$, let N_n denote the number of open paths of length n starting in 0. Since there are d^n different upward paths of length n starting at the origin, and each path has probability p^n to be open, we see that

$$\mathbb{P}[N_n \neq 0] \leq \mathbb{E}[N_n] = d^n p^n.$$

Since the events $\{N_n \neq 0\}$ decrease as $n \to \infty$ to the event $\{0 \rightsquigarrow \infty\}$, taking the limit, we see that $\mathbb{P}[0 \rightsquigarrow \infty] = 0$ for all $p < 1/d$, and therefore $1/d \leq p_c(d)$.

To prove that $p_c(d) \leq 8/9$ for $d \geq 2$, it suffices to consider the case $d = 2$, for we may view $\mathbb{Z}^2$ as a subset of $\mathbb{Z}^d$ ($d \geq 3$), and then, if there is an open path that stays in $\mathbb{Z}^2$, then certainly there is an open path in $\mathbb{Z}^d$. (Note, by the way, that in $d = 1$ one has $\mathbb{P}[0 \rightsquigarrow \infty] = 0$ for all $p < 1$, and hence $p_c(1) = 1$.)

We will use a Peierls argument, named after R. Peierls, who used a similar argument in 1936 for the Ising model [Pei36]. In Figure 7.3, we have drawn a piece of $\mathbb{Z}^2$ with a random collection of open arrows. Sites $i \in \mathbb{Z}^2$ such that $0 \rightsquigarrow i$ are drawn green. These sites are called *wet*. Consider the *dual lattice*

$$\hat{\mathbb{Z}}^2 := \{(n + \tfrac{1}{2}, m + \tfrac{1}{2}) : (n,m) \in \mathbb{Z}^2\}.$$

If there are only finitely many wet sites, then the set of all non-wet sites in $\mathbb{N}^2$ contains one infinite connected component. (Here "connected" is to be interpreted in terms of the undirected graph $\mathbb{N}^2$ with nearest-neighbor edges.) Let γ be the boundary of this infinite component. Then γ is a nearest-neighbor path in $\hat{\mathbb{Z}}^2$, starting at some point $(k + \frac{1}{2}, -\frac{1}{2})$ and ending at some point $(-\frac{1}{2}, m + \frac{1}{2})$ with $k, m \geq 0$, such that all sites immediately to the left of γ are wet, and no open arrows starting at these sites cross γ. In Figure 7.3, we have indicated γ with red arrows.

From these considerations, we see that the following statement is true: one has $0 \not\rightsquigarrow \infty$ if and only if there exists a path in $\hat{\mathbb{Z}}^2$, starting at some point $(k + \frac{1}{2}, -\frac{1}{2})$ ($k \geq 0$), ending in some point $(-\frac{1}{2}, m + \frac{1}{2})$ ($m \geq 0$), and passing to the northeast of the origin, such that all arrows of γ in the north and west directions (solid red arrows in the figure) are not crossed by an open arrow. Let M_n be the number of paths of length n with these properties. Since there are $n - 1$ dual sites from where such a path of length n can start, and since in each step, there are three directions where it can go, there are less than $n3^n$ paths of length n with

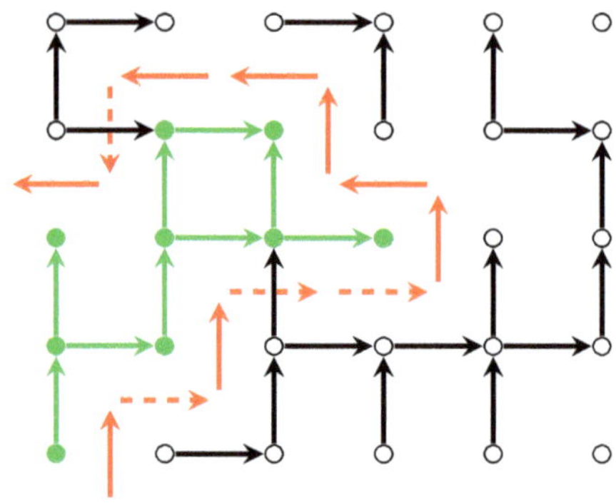

Figure 7.3 Peierls argument for oriented percolation. The green cluster of points reachable from the origin is surrounded by a red contour. The *north* and *west* steps of this contour cannot cross open arrows.

these properties. Since each path must make at least half of its steps in the north and west directions, the expected number of these paths satisfies

$$\mathbb{E}\Big[\sum_{n=2}^{\infty} M_n\Big] \leq \sum_{n=2}^{\infty} n3^n(1-p)^{n/2} < \infty \qquad (p > \tfrac{8}{9})$$

and therefore

$$\mathbb{P}[0 \not\rightsquigarrow \infty] \leq \mathbb{P}\Big[\sum_{n=2}^{\infty} M_n \geq 1\Big] \leq \mathbb{E}\Big[\sum_{n=2}^{\infty} M_n\Big] < \infty.$$

This does not quite prove what we want yet, since we need the right-hand side of this equation to be less than one. To fix this, we use a trick. (This part of the argument comes from [Dur88, Section 5a].) Set $D_m := \{0, \ldots, m\}^2$, and let us say that a set i is "wet" if $j \rightsquigarrow i$ for some $j \in D_m$. If $D_m \not\rightsquigarrow \infty$, then the set of wet sites must be finite, and, just as before, there must be a dual path surrounding this set of wet sites. Then, by the same arguments as before

$$\mathbb{P}[D_m \not\rightsquigarrow \infty] \leq \mathbb{P}\Big[\sum_{n=2m}^{\infty} M_n \geq 1\Big] \leq \mathbb{E}\Big[\sum_{n=2m}^{\infty} M_n\Big] \leq \sum_{n=2m}^{\infty} n3^n(1-p)^{n/2},$$

where now the sum starts at $2m$ since the dual paths must surround D_m and hence have length $2m$ at least. If $p > \frac{8}{9}$, then the sum is finite, so it can be made arbitrarily small by choosing m sufficiently large. It follows that $\mathbb{P}[D_m \rightsquigarrow \infty] > 0$ for some m; hence, $\mathbb{P}[i \rightsquigarrow \infty] > 0$ for at least one $i \in D_m$, and therefore, by translation invariance, also $\mathbb{P}[0 \rightsquigarrow \infty] > 0$. ∎

7.3 Survival

The main result of the present chapter is the following theorem, which rigorously establishes the existence of a phase transition for the contact process on $\mathbb{Z}^d$.

Theorem 7.3 (Nontrivial critical point) *For the nearest-neighbor or range R contact process on $\mathbb{Z}^d$ $(d \geq 1)$, the critical infection rate satisfies* $0 < \lambda_c < \infty$.

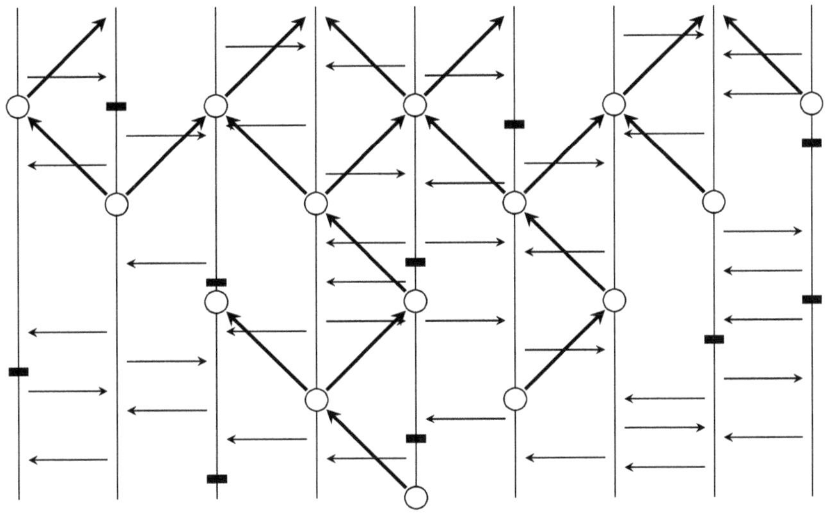

Figure 7.4 Comparison with oriented percolation. Good events in the graphical representation of the contact process correspond to open percolation arrows (thick diagonal arrows). An infinite open path along percolation arrows implies an infinite open path in the graphical representation of the contact process.

Proof As already mentioned in Section 7.1, the fact that $0 < \lambda_c$ has already been proved in Exercise 5.21. By Exercise 5.27, to prove that $\lambda_c < \infty$, it suffices to consider the one-dimensional nearest-neighbor case.

We will set up a comparison between the graphical representation of the one-dimensional nearest-neighbor contact process and oriented bond percolation on $\mathbb{Z}^2$; see Figure 7.4.

We fix $T > 0$ and define a map $\psi : \mathbb{Z}^2 \to \mathbb{Z} \times \mathbb{R}$ by

$$\psi(i) = \big(\kappa_i, \sigma_i\big) := \big(i_1 - i_2, T(i_1 + i_2)\big) \qquad \big(i = (i_1, i_2) \in \mathbb{Z}^2\big).$$

The points (κ_i, σ_i) with $i \in \mathbb{N}^2$ are indicated by open circles in Figure 7.4. As before, we make $\mathbb{Z}^2$ into a directed graph by defining a collection of arrows $\mathcal{A}$ as in (7.1). We wish to define a collection $(\omega_{(i,j)})_{(i,j)\in\mathcal{A}}$ of Bernoulli random variables such that

$$\omega_{(i,j)} = 1 \quad \text{implies} \quad (\kappa_i, \sigma_i) \rightsquigarrow (\kappa_j, \sigma_j) \qquad \big((i,j) \in \mathcal{A}\big).$$

For each $i \in \mathbb{Z}^2$ we let

$$\tau_i^{\pm} := \inf\{t \geq \sigma_i : \text{at time } t \text{ there is an infection arrow from } \kappa_i \text{ to } \kappa_i \pm 1\}$$

denote the first time after σ_i that an arrow points out of κ_i to the left or right, respectively, and we define "good events"

$$\begin{aligned} G_i^{\pm} := \big\{ & \tau_i^{\pm} < \sigma_i + T, \text{and there are no blocking symbols on} \\ & \{\kappa_i\} \times (\sigma_i, \tau_i^{\pm}] \text{ and } \{\kappa_i \pm 1\} \times (\tau_i^{\pm}, \sigma_i + T]\big\}. \end{aligned}$$

Clearly,

$$\begin{aligned} & G_i^{-} \quad \text{implies} \quad \psi(i_1, i_2) \rightsquigarrow \psi(i_1, i_2 + 1), \\ \text{and} \qquad & G_i^{+} \quad \text{implies} \quad \psi(i_1, i_2) \rightsquigarrow \psi(i_1 + 1, i_2). \end{aligned}$$

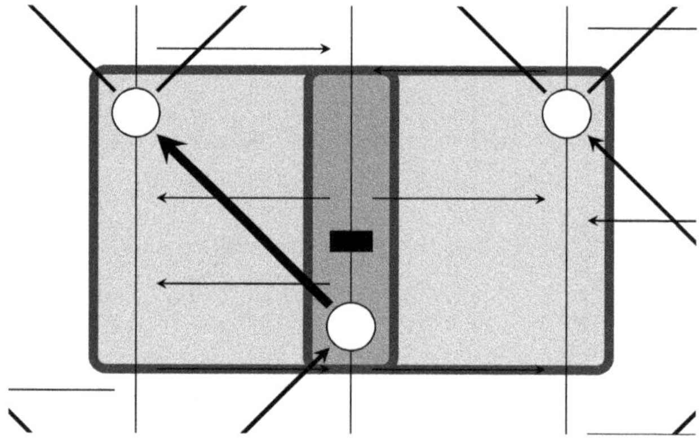

Figure 7.5 Good events use information from partially overlapping regions of space-time.

In view of this, we set

$$\omega\big((i_1,i_2),(i_1,i_2+1)\big) := 1_{\mathcal{G}_i^-} \quad \text{and} \quad \omega\big((i_1,i_2),(i_1+1,i_2)\big) := 1_{\mathcal{G}_i^+}.$$

Then the existence of an infinite open path in the oriented percolation model defined by the $(\omega_{(i,j)})_{(i,j)\in\mathcal{A}}$ implies the existence of an infinite open path in the graphical representation of the contact process, and hence survival of the latter.

We observe that

$$p := \mathbb{P}[\omega_{(i,j)} = 1] = \mathbb{P}(\mathcal{G}_i^\pm) = (1-e^{-\lambda T})e^{-T} \qquad \big((i,j)\in\mathcal{A}\big), \tag{7.2}$$

which tends to one as $\lambda \to \infty$ while $T \to 0$ in such a way that $\lambda T \to \infty$. It follows that for λ sufficiently large, by a suitable choice of T, we can make p as close to one as we wish. We would like to conclude from this that $\mathbb{P}[(0,0) \rightsquigarrow \infty] > 0$ for the oriented percolation defined by the $\omega_{(i,j)}$, and therefore also $\mathbb{P}[(0,0) \rightsquigarrow \infty] > 0$ for the contact process.

Unfortunately, life is not quite so simple, since as shown in Figure 7.5, the good events $\mathcal{G}_i^\pm$ have been defined using information from partially overlapping space-time regions of the graphical representation of the contact process, and in view of this, are not independent. They are, however, 3-dependent in the sense of Theorem 7.4 below, so by applying that result, we can estimate the Bernoulli random variables $(\omega_{(i,j)})_{(i,j)\in\mathcal{A}}$ from below by i.i.d. Bernoulli random variables $(\tilde\omega_{(i,j)})_{(i,j)\in\mathcal{A}}$ whose success probability $\tilde p$ can be made arbitrarily close to one, so we are done. ∎

7.4 K-Dependence

To finish the proof of Theorem 7.3, we need to provide the proof of Theorem 7.4, which states that K-dependent random variables with success probability p can be estimated from below by i.i.d. random variables with a success probability $\tilde p$ that tends to one as $p \to 1$.

Traditionally, for $k \geq 0$, one says that a collection $(X_i)_{i\in\mathbb{Z}^d}$ of random variables, indexed by the integer square lattice, is k-dependent if for any $A, B \subset \mathbb{Z}^d$ with

$$\inf\big\{|i-j| : i \in A,\ j \in B\big\} > k,$$

the collections of random variables $(X_i)_{i\in A}$ and $(X_j)_{j\in B}$ are independent of each other. Note that in particular, 0-dependence means independence.

It is a bit unfortunate that the traditional definition of k-dependence is strictly tied to the integer lattice $\mathbb{Z}^d$, while the structure of $\mathbb{Z}^d$ has little to do with the essential idea. Therefore, in this book, we will deviate from tradition and replace(!) the definition above by the following definition.

Let Λ be countable, and let $(X_i)_{i\in\Lambda}$ be a countable collection of random variables. Then we will say that the $(X_i)_{i\in\Lambda}$ are *K-dependent* if for each $i \in \Lambda$ there exists a $\Delta_i \subset \Lambda$ with $i \in \Delta_i$ and $|\Delta_i| \leq K$, such that

$$\chi_i \text{ is independent of } (\chi_j)_{j\in\Lambda\backslash\Delta_i}.$$

Note that according to our new definition, 1-dependence means independence. The next theorem is a reformulation of [Lig99, Thm B26], who in turn cites [LSS97].

Theorem 7.4 (K-dependence) *Let Λ be a countable set and let $p \in (0,1)$, $K < \infty$. Assume that $(\chi_i)_{i\in\Lambda}$ are K-dependent Bernoulli random variables with $\mathbb{P}[\chi_i = 1] \geq p$ $(i \in \Lambda)$, and that*

$$\tilde{p} := \left(1 - (1-p)^{1/K}\right)^2 \geq \tfrac{1}{4}.$$

Then it is possible to couple $(\chi_i)_{i\in\Lambda}$ to a collection of independent Bernoulli random variables $(\tilde{\chi}_i)_{i\in\Lambda}$ with

$$\mathbb{P}[\tilde{\chi}_i = 1] = \tilde{p} \qquad (i \in \Lambda), \tag{7.3}$$

in such a way that $\tilde{\chi}_i \leq \chi_i$ for all $i \in \Lambda$.

Proof In the language of Theorem 5.2, we must show that the law of $(\tilde{\chi}_i)_{i\in\Lambda}$ lies below the law of $(\chi_i)_{i\in\Lambda}$ in the stochastic order. Since we can always choose some arbitrary denumeration of Λ, we may assume that $\Lambda = \mathbb{N}$. Our strategy will be as follows. We will choose $\{0,1\}$-valued random variables $(\psi_i)_{i\in\Lambda}$ with $\mathbb{P}[\psi_i = 1] = r$, independent of each other and of the $(\chi_i)_{i\in\mathbb{N}}$, and put

$$\chi'_i := \psi_i\chi_i \qquad (i \in \mathbb{N}).$$

Note that the $(\chi'_i)_{i\in\mathbb{N}}$ are a thinning of the $(\chi_i)_{i\in\mathbb{N}}$. In particular, $\chi'_i \leq \chi_i$ $(i \in \mathbb{N})$, so the law of $(\chi'_i)_{i\in\Lambda}$ lies below the law of $(\chi_i)_{i\in\Lambda}$ in the stochastic order. We will show that for an appropriate choice of r,

$$\mathbb{P}[\chi'_n = 1 \,|\, \chi'_0, \ldots, \chi'_{n-1}] \geq \tilde{p} \tag{7.4}$$

for all $n \geq 0$, and we will show that this implies that the law of $(\tilde{\chi}_i)_{i\in\Lambda}$ lies below the law of $(\chi'_i)_{i\in\Lambda}$ in the stochastic order. Thus

$$\mathbb{P}\big[(\tilde{\chi}_i)_{i\in\Lambda} \in \cdot\,\big] \leq \mathbb{P}\big[(\chi'_i)_{i\in\Lambda} \in \cdot\,\big] \leq \mathbb{P}\big[(\chi_i)_{i\in\Lambda} \in \cdot\,\big],$$

which implies by Theorem 5.2 that the $(\chi_i)_{i\in\Lambda}$ can be coupled to $(\tilde{\chi}_i)_{i\in\Lambda}$ such that $\tilde{\chi}_i \leq \chi_i$ for all $i \in \Lambda$.

We start by showing that (7.4) implies that the $(\tilde{\chi}_i)_{i\in\Lambda}$ and $(\chi'_i)_{i\in\Lambda}$ can be coupled such that $\tilde{\chi}_i \leq \chi'_i$ for all $i \in \Lambda$. Set $p'_0 := \mathbb{P}[\chi'_0 = 1]$ and

$$p'_n(\varepsilon_0, \ldots, \varepsilon_{n-1}) := \mathbb{P}[\chi'_n = 1 \,|\, \chi'_0 = \varepsilon_0, \ldots, \chi'_{n-1} = \varepsilon_{n-1}]$$

whenever $\mathbb{P}[\chi'_0 = \varepsilon_0, \ldots, \chi'_{n-1} = \varepsilon_{n-1}] > 0$. Let $(U_n)_{n\in\mathbb{N}}$ be independent, uniformly distributed $[0,1]$-valued random variables. Set

$$\tilde{\chi}_n := 1_{\{U_n < \tilde{p}\}} \qquad (n \in \mathbb{N})$$

and define inductively

$$\chi'_n := 1_{\{U_n < p'_n(\chi'_0, \ldots, \chi'_{n-1})\}} \qquad (n \in \mathbb{N}).$$

Then

$$\mathbb{P}[\chi'_n = \varepsilon_n, \ldots, \chi'_0 = \varepsilon_0] = p'_n(\varepsilon_0, \ldots, \varepsilon_{n-1}) \cdots p'_1(\varepsilon_0) \cdot p'_0.$$

This shows that these new χ'_n have the same distribution as the old ones, and they are coupled to $\tilde{\chi}_i$ as in (7.3) in such a way that $\tilde{\chi}_i \leq \chi'_i$.

What makes life complicated is that (7.4) does not always hold for the original $(\chi_i)_{i\in\mathbb{N}}$, which is why we have to work with the thinned variables $(\chi'_i)_{i\in\mathbb{N}}$.[2] We observe that

$$\mathbb{P}[\chi'_n = 1 \,|\, \chi'_0 = \varepsilon_0, \ldots, \chi'_{n-1} = \varepsilon_{n-1}] = r\mathbb{P}[\chi_n = 1 \,|\, \chi'_0 = \varepsilon_0, \ldots, \chi'_{n-1} = \varepsilon_{n-1}]. \tag{7.5}$$

We will prove by induction that for an appropriate choice of r,

$$\mathbb{P}[\chi_n = 0 \,|\, \chi'_0 = \varepsilon_0, \ldots, \chi'_{n-1} = \varepsilon_{n-1}] \leq 1 - r. \tag{7.6}$$

Note that this is true for $n = 0$ provided that $r \leq p$. Let us put

$$E_0 := \{i \in \Delta_n : 0 \leq i \leq n-1,\ \varepsilon_i = 0\}, \quad E_1 := \{i \in \Delta_n : 0 \leq i \leq n-1,\ \varepsilon_i = 1\},$$
$$F := \{i \notin \Delta_n : 0 \leq i \leq n-1\}.$$

Then

$$\begin{aligned}
&\mathbb{P}[\chi_n = 0 \,|\, \chi'_0 = \varepsilon_0, \ldots, \chi'_{n-1} = \varepsilon_{n-1}] \\
&\quad = \mathbb{P}\big[\chi_n = 0 \,\big|\, \chi'_i = 0\ \forall i \in E_0,\ \chi_i = 1 = \psi_i\ \forall i \in E_1,\ \chi'_i = \varepsilon_i\ \forall i \in F\big] \\
&\quad = \mathbb{P}\big[\chi_n = 0 \,\big|\, \chi'_i = 0\ \forall i \in E_0,\ \chi_i = 1\ \forall i \in E_1,\ \chi'_i = \varepsilon_i\ \forall i \in F\big] \\
&\quad = \frac{\mathbb{P}\big[\chi_n = 0,\ \chi'_i = 0\ \forall i \in E_0,\ \chi_i = 1\ \forall i \in E_1,\ \chi'_i = \varepsilon_i\ \forall i \in F\big]}{\mathbb{P}\big[\chi'_i = 0\ \forall i \in E_0,\ \chi_i = 1\ \forall i \in E_1,\ \chi'_i = \varepsilon_i\ \forall i \in F\big]} \\
&\quad \leq \frac{\mathbb{P}\big[\chi_n = 0,\ \chi'_i = \varepsilon_i\ \forall i \in F\big]}{\mathbb{P}\big[\psi_i = 0\ \forall i \in E_0,\ \chi_i = 1\ \forall i \in E_1,\ \chi'_i = \varepsilon_i\ \forall i \in F\big]} \\
&\quad = \frac{\mathbb{P}\big[\chi_n = 0 \,\big|\, \chi'_i = \varepsilon_i\ \forall i \in F\big]}{\mathbb{P}\big[\psi_i = 0\ \forall i \in E_0,\ \chi_i = 1\ \forall i \in E_1 \,\big|\, \chi'_i = \varepsilon_i\ \forall i \in F\big]} \\
&\quad \leq \frac{1-p}{(1-r)^{|E_0|}\mathbb{P}\big[\chi_i = 1\ \forall i \in E_1 \,\big|\, \chi'_i = \varepsilon_i\ \forall i \in F\big]} \leq \frac{1-p}{(1-r)^{|E_0|}\, r^{|E_1|}},
\end{aligned} \tag{7.7}$$

where in the last step we have used K-dependence and the (nontrivial) fact that

$$\mathbb{P}\big[\chi_i = 1\ \forall i \in E_1 \,\big|\, \chi'_i = \varepsilon_i\ \forall i \in F\big] \geq r^{|E_1|}. \tag{7.8}$$

[2] Indeed, let $(\phi_n)_{n\geq 0}$ be independent $\{0,1\}$-valued random variables with $\mathbb{P}[\phi_n = 1] = \sqrt{p}$ for some $p < 1$, and put $\chi_n := \phi_n \phi_{n+1}$. Then the $(\chi_n)_{n\geq 0}$ are 3-dependent with $\mathbb{P}[\chi_n = 1] = p$, but $\mathbb{P}[\chi_n = 1 | \chi_{n-1} = 0, \chi_{n-2} = 1] = 0$.

We claim that (7.8) is a consequence of the induction hypothesis (7.6). Indeed, we may assume that the induction hypothesis (7.6) holds regardless of the ordering of the first n elements, so without loss of generality, we may assume that $E_1 = \{n-1, \dots, m\}$ and $F = \{m-1, \dots, 0\}$, for some m. Then the left-hand side of (7.8) may be written as

$$\prod_{k=m}^{n-1} \mathbb{P}\big[\chi_k = 1 \,\big|\, \chi_i = 1 \ \forall m \le i < k,\ \chi'_i = \varepsilon_i \ \forall 0 \le i < m\big]$$
$$= \prod_{k=m}^{n-1} \mathbb{P}\big[\chi_k = 1 \,\big|\, \chi'_i = 1 \ \forall m \le i < k,\ \chi'_i = \varepsilon_i \ \forall 0 \le i < m\big] \ge r^{n-m}.$$

If we assume, moreover, that $r \ge \frac{1}{2}$, then $r^{|E_1|} \ge (1-r)^{|E_1|}$ and therefore the right-hand side of (7.7) can be further estimated as

$$\frac{1-p}{(1-r)^{|E_0|}\, r^{|E_1|}} \le \frac{1-p}{(1-r)^{|\Delta_n \cap \{0,\dots,n-1\}|}} \le \frac{1-p}{(1-r)^{K-1}}.$$

We see that in order for our proof to work, we need $\frac{1}{2} \le r \le p$ and

$$\frac{1-p}{(1-r)^{K-1}} \le 1 - r. \tag{7.9}$$

In particular, choosing $r = 1 - (1-p)^{1/K}$ yields equality in (7.9). Having proved (7.6), we see by (7.5) that (7.4) holds provided that we put $\tilde{p} := r^2$. ∎

Exercise 7.5 *Combine Theorem 7.2 and formulas (7.2) and (7.3) to derive an explicit upper bound on the critical infection rate λ_c of the one-dimensional contact process.*

Exercise 7.6 *The one-dimensional contact process with double deaths has been introduced just before Exercise 5.24. Use comparison with oriented percolation to prove that the one-dimensional contact process with double deaths survives with positive probability if its branching rate λ is large enough. When you apply Theorem 7.4, what value of K do you (at least) need to use?*

Exercise 7.7 *Use the previous exercise and Exercise 5.24 to conclude that for the cooperative branching process considered there, if λ is large enough, then: (i) If the process is started with at least two particles on neighboring sites, then there is a positive probability that there will always be pairs of particles on neighboring sites. (ii) The upper invariant law is nontrivial.*

Exercise 7.8 *Assume that there exists some $t > 0$ such that the contact process satisfies*

$$r := \mathbb{E}^{e_0}\big[|X_t|\big] < 1.$$

Show that this then implies that

$$\mathbb{E}^{e_0}\big[|X_{nt}|\big] \le r^n \qquad (n \ge 0)$$

and the process started in any finite initial state dies out a.s. Can you use this to improve the lower bound $1/|\mathcal{N}_i| \le \lambda_c$ from Exercise 5.21, for example, for the one-dimensional nearest-neighbor process?

References

[AB05] D. J. Aldous and A. Bandyopadhyay. A survey of max-type recursive distributional equations. *Ann. Appl. Probab.* 15(2) (2005), 1047–1110.

[ADS15] M. Aizenman, H. Duminil-Copin, and V. Sidoravicius. Random currents and continuity of Ising model's spontaneous magnetization. *Commun. Math. Phys.* 334(2) (2015), 719–742.

[BD01] W. A. Brock and S. N. Durlauf. Discrete choice with social interactions. *Rev. Econ. Stud.* 68(2) (2001), 235–260.

[BDD91] M. Bramson, W. Ding, and R. Durrett. Annihilating branching processes. *Stoch. Process. Appl.* 37(1) (1991), 1–17,

[BDT19] O. Blondel, A. Deshayes, and C. Toninelli. Front evolution of the Fredrickson-Andersen one spin facilitated model. *Electron. J. Probab.* 24 (2019), Paper No. 1, 1–32.

[BG90] C. Bezuidenhout and G. Grimmett. The critical contact process dies out. *Ann. Probab.* 18(4) (1990), 1462–1482.

[CD+20] R. Cerf, P. Dai Pra, M. Formentin, and D. Tovazzi. Rhythmic behavior of an Ising model with dissipation at low temperature. *ALEA, Lat. Am. J. Probab. Math. Stat.* 18(1) (2021), 439–467.

[CFT16] F. Collet, M. Formentin, and D. Tovazzi. Rhythmic behavior in a two-population mean-field Ising model. *Phys. Rev. E* 94 (2016), 042139.

[Cox84] J. T. Cox. An alternate proof of a correlation inequality of Harris. *Ann. Probab.* 12 (1984), 272–273.

[CP14] J. T. Cox and E. A. Perkins. A complete convergence theorem for voter model perturbations. *Ann. Appl. Probab.* 24(1) (2014), 150–197.

[CS73] P. Clifford and A. Sudbury. A model for spatial conflict. *Biometrika* 60(3) (1973), 581–588.

[DFR13] P. Dai Pra, M. Fischer, and D. Regoli. A Curie-Weiss model with dissipation. *J. Stat. Phys.* 152(1) (2013), 37–53.

[DLZ14] R. Durrett, T. Liggett, and Y. Zhang. The contact process with fast voting. *Electron. J. Probab.* 19 (2014), Paper No. 28, 1–19.

[DN08] R. W. R. Darling and J. R. Norris. Differential equation approximations for Markov chains. *Probab. Surv.* 5 (2008), 37–79.

[Dob71] R. L. Dobrushin. Markov processes with a large number of locally interacting components: Existence of a limit process and its ergodicity. *Problems Inform. Transm.* 7(2) (1971), 149–164.

[DST17] H. Duminil-Copin, V. Sidoravicius, and V. Tassion. Continuity of the phase transition for planar random-cluster and Potts models with $1 \leq q \leq 4$. *Commun. Math. Phys.* 349(1) (2017), 47–107.

[Dur88] R. Durrett. *Lecture Notes on Particle Systems and Percolation.* Wadsworth & Brooks/Cole, Pacific Grove, 1988.

[Dur91] R. Durrett. A new method for proving the existence of phase transitions. Pages 141–169 in: K. S. Alexander and J. C. Watkins (eds). *Spatial Stochastic Processes: A Festschrift. in Honor of Ted Harris on His Seventieth Birthday.* Progress in Probability 19, Birkhäuser, Boston, 1991.

[Dur92] R. Durrett. Stochastic growth models: Bounds on critical values. *J. Appl. Probab.* 29(1) (1992), 11–20.

[Dur95] R. Durrett. Ten lectures on particle systems. Pages 97–201 in: P. Biane (ed.) *Lectures on Probability Theory.* École d'été de probabilités de Saint-Flour XXIII – 1993. Springer, Berlin, 1995.

[EK86] S. N. Ethier and T. G. Kurtz. *Markov Processes: Characterization and Convergence.* John Wiley & Sons, New York, 1986.

[Fer13] P. L. Ferrari. *Why Random Matrices Share Universal Processes with Interacting Particle Systems?* Lecture notes, 2013, arXiv:1312.1126.

[FKG71] C. M. Fortuin, P. W. Kasteleyn, and J. Ginibre. Correlation inequalities on some partially ordered sets. *Commun. Math. Phys.* 22 (1971), 89–103.

[FM01] J. A. Fill and M. Machida. Stochastic monotonicity and realizable monotonicity. *Ann. Probab.* 29(2), (2001), 938–978.

[Gla63] R. J. Glauber. Time-dependent statistics of the Ising model. *J. Math. Phys.* 4 (1963), 294–307.

[GB07] T. Gobron and I. Merola. First-order phase transition in Potts models with finite-range interactions. *J. Stat. Phys.* 126(3) (2007), 507–583.

[GR25] C. Giardinà and F. Redig *Duality for Markov Processes: A Lie Algebraic Approach.* Springer, Cham, 2025.

[GK+09] C. Giardinà, J. Kurchan, F. Redig, and K. Vafayi. Duality and hidden symmetries in interacting particle systems. *J. Stat. Phys.* 135(1) (2009), 25–55.

[GP+18] B. Garrod, M. Poplavskyi, R. P. Tribe, and O. V. Zaboronski. Examples of interacting particle systems on $\mathbb{Z}$ as Pfaffian point processes: Annihilating and coalescing random walks. *Ann. Henri Poincaré* 19(12) (2018), 3635–3662.

[GPA01] M. Ghaemi, G. A. Parsafar, and M. Ashrafizaadeh. Calculation of the critical temperature for 2- and 3-dimensional Ising models and for 2-dimensional Potts models using the transfer matrix method. *J. Phys. Chem. B* 105(42) (2001), 10355–10359.

[Gri79] D. Griffeath. *Additive and Cancellative Interacting Particle Systems.* Lecture Notes in Math. 724, Springer, Berlin, 1979.

[Gri18] G. Grimmett. *Probability on Graphs. Random Processes on Graphs and Lattices. 2nd Edition.* Institute of Mathematical Statistics Textbooks 8. Cambridge University Press, Cambridge, 2018.

[Han99] S. J. Handjani. The complete convergence theorem for coexistent threshold voter models. *Ann. Probab.* 27(1) (1999) 226–245.

[Har72] T. E. Harris. Nearest-neighbor Markov interaction processes on multidimensional lattices. *Adv. Math.* 9 (1972), 66–89.

[Har74] T. E. Harris. Contact interactions on a lattice. *Ann. Probab.* 2 (1974), 969–988.

[Har76] T. E. Harris. On a class of set-valued Markov processes. *Ann. Probab.* 4 (1976), 175–194.

[Har77] T. E. Harris. A correlation inequality for Markov processes in partially ordered state spaces. *Ann. Probab.* 5 (1977), 451–454.

[Har78] T. E. Harris. Additive set-valued Markov processes and graphical methods. *Ann. Probab.* 6 (1978), 355–378.

[HHL08] M. Henkel, H. Hinrichsen, and S. Lübeck. *Non-Equilibrium Phase Transitions. Vol. 1: Absorbing Phase Transitions.* Springer, Dordrecht, 2008.

[HT25] I. Hartarsky and C. Toninelli. *Kinetically Constrained Models.* Springer, Berlin, 2025.

[Isi25] E. Ising. Beitrag zur Theorie des Ferromagnetismus. *Z. Phys.* 31 (1925), 253–258.

[JK14a] B. Jahnel and C. Külske. A class of nonergodic interacting particle systems with unique invariant measure. *Ann. Appl. Probab.* 24(6) (2014), 2595–2643.

[JK14b] S. Jansen and N. Kurt. On the notion(s) of duality for Markov processes. *Prob. Surv.* 11 (2014), 59–120.

[Jou07] A. Joulin. Poisson-type deviation inequalities for curved continuous-time Markov chains. *Bernoulli* 13(3) (2007), 782–798.

[Kal97] O. Kallenberg. *Foundations of Modern Probability.* Springer, New York, 1997.

[KKO77] T. Kamae, U. Krengel, and G. L. O'Brien. Stochastic inequalities on partially ordered spaces. *Ann. Probab.* 5 (1977), 899–912.

[Kro99] S. Krone. The two-stage contact process. *Ann. Appl. Probab.* 9(2) (1999), 331–351.

[Lan24] N. Lanchier. *Stochastic Interacting Systems in Life and Social Sciences.* Walter de Gruyter, Berlin, 2024.

[Lat24] J. N. Latz. *Pathwise Duality of Interacting Particle Systems.* PhD thesis, Faculty of Mathematics and Physics, Charles University, Prague, 2024. `https://library.utia.cas.cz/separaty/2024/SI/latz-0598658.pdf`

[Led00] M. Ledoux. The geometry of Markov diffusion generators. *Ann. Fac. Sci. Toulouse, VI. Sér., Math.* 9(2) (2000), 305–366.

[Lig72] T. M. Liggett. Existence theorems for infinite particle systems. *Trans. Am. Math. Soc.* 165 (1972), 471–481.

[Lig85] T. M. Liggett. *Interacting Particle Systems.* Springer-Verlag, New York, 1985.

[Lig95] T. M. Liggett. Improved upper bounds for the contact process critical value. *Ann. Probab.* 23(2) (1995), 697–723.

[Lig99] T. M. Liggett. *Stochastic Interacting Systems: Contact, Voter and Exclusion Processes.* Springer-Verlag, Berlin, 1999.

[Lig10] T. M. Liggett. *Continuous time Markov Processes. An Introduction.* AMS, Providence, 2010.

[LS21] J. N. Latz and J. M. Swart. Commutative monoid duality. *J. Theor. Probab.* 36(2) (2023), 1088–1115.

[LS23a] J. N. Latz and J. M. Swart. Applying monoid duality to a double contact process. *Electron. J. Probab.* 28 (2023), paper no. 70, 1–26.

[LSS97] T. M. Liggett, R. H. Schonmann and A. M. Stacey. Domination by product measures. *Ann. Probab.* 25(1) (1997), 71–95.

[Loe63] M. Loève. *Probability Theory, 3rd edition.* Van Nostrand, Princeton, 1963.

[Mar99] F. Martinelli. Lectures on Glauber dynamics for discrete spin models. Pages 93–191 in: P. Bernard (ed.) *Lectures on Probability Theory and Statistics.* Lecture Notes in Mathematics 1717. Springer, Berlin, 1999.

[Mou95] T. S. Mountford. A coupling of infinite particle systems. *J. Math. Kyoto Univ.* 35(1) (1995), 43–52.

[MQ17] K. Matetski and J. Quastel. *From the Totally Asymmetric Simple Exclusion Process to the KPZ Fixed Point.* Lecture Notes, 2017, arXiv:1710.02635.

[MSS20] T. Mach, A. Sturm, and J. M. Swart. Recursive tree processes and the mean-field limit of stochastic flows. *Electron. J. Probab.* 25 (2020), paper No. 61, 1–63.

[MW66] N. D. Mermin and H. Wagner. Absence of ferromagnetism or antiferromagnetism in one- or two-dimensional isotropic Heisenberg models. *Phys. Rev. Lett.* 17(22) (1966), 1133–1136.

[Nac65] L. Nachbin. *Topology and Order*. Princeton, New Jersey, 1965.

[Neu94] C. Neuhauser. A long range sexual reproduction process. *Stoch. Process. Appl.* 53(2) (1994), 193–220.

[Nob92] C. Noble. Equilibrium behavior of the sexual reproduction process with rapid diffusion. *Ann. Probab.* 20(2) (1992), 724–745.

[NP99] C. Neuhauser and S. W. Pacala. An explicitly spatial version of the Lotka-Volterra model with interspecific competition. *Ann. Appl. Probab.* 9(4) (1999), 1226–1259.

[Ons44] L. Onsager. Crystal statistics, I. A two-dimensional model with an order-disorder transition. *Phys. Rev.*, II Ser. 65 (1944), 117–149.

[Pei36] R. Peierls. On Ising's model of ferromagnetism. *Proc. Cambridge Philos. Soc.* 32 (1936), 477–481.

[Pen08] M. D. Penrose. Existence and spatial limit theorems for lattice and continuum particle systems. *Probab. Surv.* 5 (2008), 1–36.

[Pre74] C. J. Preston. A generalization of the FKG inequalities. *Comm. Math. Phys.* 36(3) (1974), 233–241.

[RS03] F. Ritort and P. Sollich. Glassy dynamics of kinetically constrained models. *Adv. Phys.* 52(4) (2003), 219–342.

[Sal97] L. Saloff-Coste. Lectures on finite Markov chains. Pages 301–413 in: P. Bernard (eds) *Lectures on Probability Theory and Statistics.* Lecture Notes in Mathematics 1665. Springer, Berlin, 1997.

[SL95] A. Sudbury and P. Lloyd. Quantum operators in classical probability theory. II: The concept of duality in interacting particle systems. *Ann. Probab.* 23(4) (1995), 1816–1830.

[SL97] A. Sudbury and P. Lloyd. Quantum operators in classical probability theory. IV: Quasi-duality and thinnings of interacting particle systems. *Ann. Probab.* 25(1) (1997), 96–114.

[Spi70] F. Spitzer. Interaction of Markov processes. *Adv. Math.* 5 (1970), 246–290.

[SS08] A. Sturm and J. M. Swart. Voter models with heterozygosity selection. *Ann. Appl. Probab.* 18(1) (2008), 59–99.

[SS15] A. Sturm and J. M. Swart. A particle system with cooperative branching and coalescence. *Ann. Appl. Probab.* 25(3) (2015), 1616–1649.

[SS18] A. Sturm and J. M. Swart. Monotone and additive Markov process duality. *J. Theor. Probab.* 31(2) (2018), 932–983.

[Sud97] A. Sudbury. The convergence of the biased annihilating branching process and the double-flipping process in $\mathbb{Z}^d$. *Stoch. Proc. Appl.* 68(2) (1997), 25–264.

[Sud99] A. Sudbury. Hunting submartingales in the jumping voter model and the biased annihilating branching process. *Adv. Appl. Probab.* 31(3) (1999), 839–854.

[Sud00] A. Sudbury. Dual families of interacting particle systems on graphs. *J. Theor. Probab.* 13(3) (2000), 695–716.

[Swa06] J. M. Swart. Duals and thinnings of some relatives of the contact process. Preprint (2006), arXiv:math.PR/0604335.

[Tov19] D. Tovazzi. *Self-sustained periodic behaviors in interacting systems: Macroscopic limits and fluctuations.* PhD thesis, University of Padua, 2019.

[Vas69] N. B. Vasil'ev. Limit behavior of one random medium. (In Russian.) *Probl. Peredachi Inf.* 5(4) (1969) 68–74.

Index

For EU product safety concerns, contact us at Calle de José Abascal, 56–1°,
28003 Madrid, Spain or eugpsr@cambridge.org.

www.ingramcontent.com/pod-product-compliance
Ingram Content Group UK Ltd.
Pitfield, Milton Keynes, MK11 3LW, UK
UKHW051935150726
473310UK00005B/193

* 9 7 8 1 0 0 9 8 4 3 4 7 8 *